The Red Polled Cattle Herdbook
Cattle Descended from the Norfolk and Suffolk Red Polled Cattle

by Henry F. Euren

with an introduction by Jackson Chambers

This work contains material that was originally published in 1883.

This publication is within the Public Domain.

This edition is reprinted for educational purposes
and in accordance with all applicable Federal Laws.

Introduction Copyright 2017 by Jackson Chambers

Self Reliance Books

Get more historic titles on animal and stock breeding, gardening and old fashioned skills by visiting us at:

http://selfreliancebooks.blogspot.com/

Introduction

I am pleased to present another title in the "Cattle" series.

The work is in the Public Domain and is re-printed here in accordance with Federal Laws.

As with all reprinted books of this age that are intended to perfectly reproduce the original edition, considerable pains and effort had to be undertaken to correct fading and sometimes outright damage to existing proofs of this title. At times, this task is quite monumental, requiring an almost total "rebuilding" of some pages from digital proofs of multiple copies. Despite this, imperfections still sometimes exist in the final proof and may detract from the visual appearance of the text.

I hope you enjoy reading this book as much as I enjoyed making it available to readers again.

Jackson Chambers

DOLLY — N 2.
Register Number 1433.

THE PROPERTY OF AND BRED BY J. J. COLMAN, ESQ., M.P.,

Calved November 3, 1879; produced her first Calf November 10, 1882.

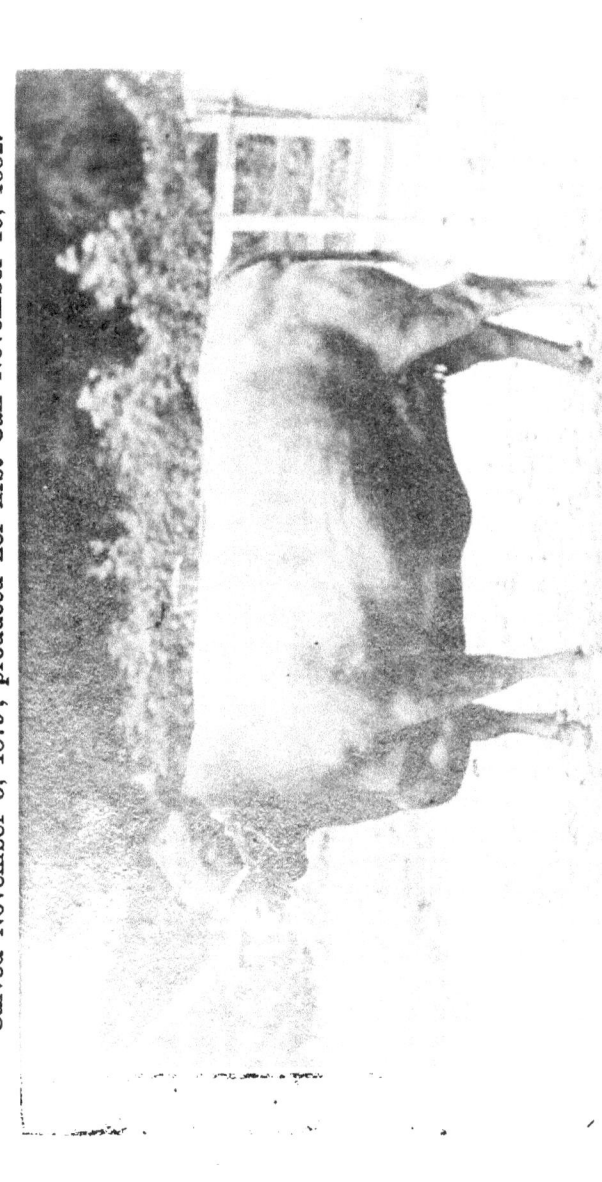

HONOURS WON:—1st Royal, 1st Norfolk, 1st Suffolk, 1881. 1st Norfolk and President's Cup offered for Best Cow or Heifer, 2nd Suffolk, 1882.

Photographed from Life by Messrs. Mann and Adcock, Norwich, August, 1882.

SUBSCRIBERS.

Her Grace THE DUCHESS OF HAMILTON and BRANDON, Easton Park, Suffolk.

The Most Hon. THE MARQUIS OF BRISTOL, Ickworth Park, Suffolk.

The Right Hon. THE EARL OF KIMBERLEY, Kimberley House, Norfolk.

The Right Hon. LORD HASTINGS, Melton Constable, Norfolk.

The Right Hon. LORD HENNIKER, Thornham Hall, Suffolk.

The Right Hon. LORD SUFFIELD, K.C.B., Gunton, Norfolk.

Sir ROBERT J. BUXTON, Bart., M.P., Shadwell Court, Norfolk.

Sir EDWARD C. KERRISON, Bart., Oakley Park, Suffolk.

SMITH, The Right Hon. W. H., M.P., Great Thurlow, Suffolk.

ALLEN, Rev. S., D.D., Shouldham Hall, Norfolk.

AMHERST, W. A. T., Esq., M.P., Didlington Hall, Norfolk.

AUSTIN, C., Esq., Brandeston Hall, Suffolk.

BAKER, Mr. John, Colville House, Wisbech, Cambs.

BALY, Mr. John, Hardingham, Norfolk.

BIDDELL, Mr. HERMAN, Playford, Suffolk.

BIRKBECK, HENRY, Esq., Stoke Holy Cross, Norfolk.

BLOFIELD, Mr. PETER, Quidenham, Norfolk.

BOGGIS, Mr. JOSEPH, Jun., Geldeston, Norfolk.

BOON, Mr. E., Executors of, Brandeston, Suffolk.

BRADFIELD, Mr. W., Ramsley Farm, Elmham, Norfolk.

BROWN, Mr. THOMAS, Marham Hall, Norfolk.

BUSK, Mr. W. G., Coombe Bissett, Salisbury, Wilts.

SUBSCRIBERS.

CARMALT, Mr. JAS. E., Scranton, Pennsylvania, U.S.A.

CHAMBERLIN, Mr. H. J., Davilla, Milan Co., Texas, U.S.A.

CLARK, Messrs. J. H. and W. W., Toledo, Washington Co., Pennsylvania, U.S.A.

COLLYER, Mrs., Hackford Hall, Reepham, Norfolk.

COLMAN, J. J., Esq., M.P., Carrow House, Norwich, Norfolk.

CORDY, Mr. C. K., Trimley St. Mary, Suffolk.

DOUGLAS, Mr. S. B., Reed's Corner, Ontario Co., New York, U.S.A.

EASTER, Mr. W. B., Stockton Hall, Bungay, Norfolk.

ENGLISH, Mr. E. W., Saranac, Michigan, U.S.A.

FORRESTER, GEORGE, Esq., Norwich, Norfolk.

FULCHER, Mr. T., Elmham, Norfolk.

GELDARD, Mr. ROBERT, Cappleside, Settle, Yorkshire.

GIRLING, Rev. J. C., Hautbois Rectory, Norfolk.

GOODERHAM, Mr. G., Monewden, Suffolk.

HAGGARD, BAZETT M., Esq., Kirby Cane Hall, Norfolk.

HAMMOND, Mr. JOHN, Bale, Norfolk.

HARRISON, T., Esq., Copford Hall, Essex.

HARTOPP, Sir JOHN W. C., Bart., Kingswood Warren, Epsom, Surrey.

HARVEY, Mr. W., Timworth, Suffolk.

HAYLOCK, Mr. H., Saham Grove, Norfolk.

HOLMES, GEORGE, Esq., Brooke, Norfolk.

HOWELL, Mr. JOHN, Great Walsingham, Norfolk.

HUDSON, Mr. WILLIAM, Quarles, Holkham, Norfolk.

JONES, Messrs. D. and G., Galesburg, Illinois, U.S.A.

KENT, Mr. F. D., Corringham, Essex.

KIMBALL, R. J., Esq., 18, Wall Street, New York.

KNAPP, Mr. J. M., Bellevue, Michigan, U.S.A.

LEGGE, Rev. A. G., Elmham, Norfolk.

SUBSCRIBERS.

LE STRANGE, H. Esq., Hunstanton Hall, Norfolk.

LEWIS, Mr. J. T., Chatham, Sangamon Co., Illinois, U.S.A.

LOFFT, R. E., Esq., Troston Hall, Suffolk.

LOMBE, Rev. H. EVANS, Bylaugh Hall, Norfolk.

LONG, Lieut.-Col. W. BEESTON, Hurts Hall, Saxmundham, Suffolk.

LUBBOCK, J. J. L., Esq., Catfield Hall, Norfolk.

MANN, Mr. FAIRMAN J., Shropham Hall, Norfolk.

MARGARSON, Mr. J., Wendling, Norfolk.

MARRIOTT, J. L., Esq., Narborough, Norfolk.

MASON, R. HARVEY, Esq., Necton Hall, Norfolk.

MEAD and KIMBALL, Messrs., Suffolk Farm, Randolph, Vermont, U.S.A.

MORRIS, Mr. FREDK., Geldeston, Norfolk.

MUSTARD, Mr. J. L., Lebanon, Missouri, U.S.A.

NATHUSIUS, Herr A. E. von, Meyendorf, Magdeburg, Germany.

NORFOLK AGRICULTURAL ASSOCIATION.

PAINE, Mr. G. J., Risby, Suffolk.

PALMER, Mr. T. L., Banham, Norfolk.

PERKINS, Mrs., Saham Hall, Norfolk.

POWELL, Mr. NICHOLAS, Glandford, Norfolk.

READ, C. S., Esq., Honingham Thorpe, Norfolk.

RIVETT, Mr. J., Mileham, Norfolk.

ROGERS, J. F., Esq., Swanington, Norfolk.

Ross, General L. F., Iowa City, Iowa, U.S.A.

ROYAL AGRICULTURAL SOCIETY.

SANDERSON, G. L., Williamsport, Pennsylvania, U.S.A.

SMITH, Mr. A. J., Rendlesham, Suffolk.

SMITH, Mr. J. Mc LAIN, Dayton, Ohio, U.S.A.

SPINKS, Mr. J. M., Harpley, Norfolk.

SPURLING, Mr. H., Shotley, Suffolk.

STEVENS, Mr. D. L., Elkdale, Susquehanna Co., Pennsylvania, U.S.A.

STIMPSON, Mr. B., Morton-on-the-Hill, Norfolk.

TABER, Mr. G. F., Ravinewood Farm, Patterson, New York, U.S.A.

TABER, Mr. G. K., Pawling, New York, U.S.A.

TAYLOR, ALFRED, Esq., Starston Place, Norfolk.

TAYLOR, Mr. GARRETT, Trowse House, Norwich.

TIBBITS, Capt. J. BORLASE, Barton Seagrave, Northamptonshire.

WARING, HENRY, Esq., Beenham House, Reading, Berks.

WATERS, Mr. C., Postwick Grange, Norfolk.

WEST, Mr. T. R., Raveningham, Norfolk.

WOLTON, Mr. H., Newbourn Hall, Suffolk.

WOLTON, Mr. S., Butley Abbey, Suffolk.

INTRODUCTION.

Early in January the Editor issued a circular letter to owners of Registered Red Polled Cattle, in which he said:—

The Council of the Royal Agricultural Society of England has just sanctioned a rule under which, henceforth, all Cattle exhibited at the Society's Shows must either be already registered or be qualified for registration in the Herd Book of the respective breeds. This fact has led to the consideration of the future course which will best serve the interests of the breeders of Red Polled Stock, and also help to make registration more valuable. I have arrived at the conclusion that both these ends will be most efficiently met in the following way, viz.:—To issue in the spring of *every alternate year* an alphabetical register of Bulls dropped before January 1st, in use, intended to be used, or to be offered for sale, of Cows not previously registered, and of Heifers dropped before January 1st preceding the date of issue.

Various expedients have been proposed by the supporters of Herd Books of other breeds to secure a complete record of the produce of each Cow, without also adding so largely to the bulk of a printed volume as must be the result if this record cover only a very limited period. It appears to me that all that is required may be done, if we make it a rule to add this record as a supplementary sheet, say at intervals of ten years.

A proposal I have made, that we adopt, as the recognised public name of registered stock, the simple title "Red Polled" has been most favourably received. The change is fully warranted by the fact that the cattle are no longer confined to the two counties of Norfolk and Suffolk. I propose, therefore, to make the title of Vol. 2, read as follows:—"THE RED POLLED HERD BOOK, of Cattle descended from the Norfolk and Suffolk Red Polled."

The Conditions of Registry are these:—

(a) That the Cattle are the progeny of sire and dam of registered stock, or of stock entitled by descent to be registered: or

(b) That they are descended in direct line as regards both sire and dam during three generations, from cattle registered in the Foundation Volume—the Norfolk and Suffolk Red Polled Herd Book: or

(c) That there is good and satisfactory evidence of descent from an old established herd in Norfolk or Suffolk, and that the sires have for at least five generations been Red Polled Stock answering to the "Essentials" of The Standard Description: or

INTRODUCTION.

> (*d*) That the sire, grand sire, and dam of the Cow or Heifer to be registered have answered to the "Essentials" of the "Standard Description." The male progeny of these last named "Probationary" Cattle are not recognised as of pure blood. Descendants of Probationers cannot be registered as pure until after five generations have been recorded.
>
> Cattle not tracing their descent from Stock already registered, and Probationers, will henceforth be grouped thus: 1 Norf., 2 Norf., &c., 1 Suff., 2 Suff., &c., or, if Graded Stock, the produce of Red Polled Bulls in America, 1 N. Y., 2 N. Y., &c., 1 Virg., 2 Virg., &c., 1 Ill., 2 Ill., &c., &c.

I have to acknowledge with thanks the kindness of several gentlemen who sent me returns of milk and butter yields, live and dead weights, &c., thus enabling me to prepare for the *Live Stock Journal Almanack* for 1883 an article on Red Polled Cattle, containing such facts as will permit of comparison with other breeds. The article is illustrated with portraits of stock from photographs, and may thus help to make known on the Continent, in America, and in Australia, the excellencies of Red Polled Cattle. Similar good service has been rendered by Mr. G. F. Taber, of Patterson, New York; and Mr. R. J. Kimball, of New York City, with his partner Col. Mead, of Suffolk Farm, Randolph Co., Vermont, who, as importers of the cattle, have published in the Agricultural Papers of the United States excellent portraits of Red Polls.

The two years which have elapsed since the issue of the First Part of this volume have been full of promise for the breeders of Red Polled Cattle: prices have advanced considerably, and the merits of the stock have been tested over a considerably enlarged area.

As will have been learned from the circular above quoted, Registered Stock descended from the Norfolk and Suffolk Red Polled are henceforth to be known as "Red Polled." The Editor would strongly urge on Breeders the importance of great accuracy in keeping the Herd Register, so that there may be no necessity for future amendment of a pedigree when once it has been set forth in the Herd Book. Errors in transcription may occur, and, as will be seen to be the case in this issue, a "Corrected Entry" may be inserted at the first opportunity which presents itself; but this is a defect which additional care on the part of the Breeder may altogether remove. The Editor finds that the plan of registry he has now adopted has greatly reduced the chance of error in name and numbers when putting the MS. entries into type.

By the kind permission of Messrs. Cassell and Co. (Limited), the Editor is enabled to reproduce the article spoken of in the circular, with important additions. Copies of this Essay, for general circulation

INTRODUCTION.

among persons asking for information about Red Polls, can be had at a small cost.

The Editor is agreeably surprised at the large number of entries made for the Second Part of Vol. II., and believes the issues of 1881 and 1883, when bound together, will make as large a book as can be conveniently handled. He, therefore, begs to announce that the next issue—in the spring of the year 1885—will be the First Part of Vol. III., entries for which should be made early in January of that year.

The experience which the Editor has gained has led him to the conclusion that every published pedigree of a cow should extend to four generations. This is the rule which has been followed in this issue, where four or more generations have been recorded. In other particulars there has been no alteration of the plans followed in the 1881 issue.

Under the provisions of one of the Conditions of Registry, there are, in this issue, six new

FOUNDATION TRIBES.

1 Norfolk : POND.

In the Foundation Volume, mention was made of the "old and very good herd of Mr. B. Pond, of Dunham," as having been a source of supply both for the Elmham Herd and for Mr. Powell's Herd. This strain of Red Polled Stock, before the Herd Book had been founded, passed into the possession of Mr. W. Wiffen, of Tittleshall. He regularly, and as far as the Editor can learn, without exception, resorted to Elmham or to Ramsley for Red Polled Bulls. The Stock thus bred are now grouped under the one tribal name "Pond."

2 Norfolk : MANN.

An old strain of blood-red Polled stock was kept for fifty years by Mrs. Mary Mann, of Great Ellingham. In 1862 Mrs. Mann's nephew, Mr. F. J. Mann, bought two cows which were bred by his aunt, or were descended from this old Norfolk Red Polled herd. In the following year he bought, at East Dereham, two heifers, bred and offered for sale by Lord Sondes. From 1862 to 1866 the bulls used were bred by Mrs. Mann, or from her stock. From 1866 to 1871, bulls were used bred from the Elmham heifers. In 1869 the herd was increased by the purchase of a Red Polled cow, bred by Mr. Applewhaite, of Pickenham. The calf, by a Pickenham bull, which was produced by this cow, was used from 1871, and since that time the bulls used have all been descended from the Applewhaite cow. All the cows now registered by Mr. Mann are thus of similar blood, and are formed into one tribe.

3 Norfolk : NICHOLSON.

One of the Elmham tenants, Mr. J. Nicholson, farmed at Gressenhall, and was a breeder of Red Polled stock for many years. Redjacket 3rd 165 appears in the Herd Book as of his breeding. He either used bulls bred on his farm, or those bred at Elmham. The cows now registered in this tribe were purchased when he retired from farming at Michaelmas, 1881.

4 Norfolk : PECK.

This tribe is descended from a cow bought many years ago at East Dereham. Her descendants have been sired by Elmham or Wilby bulls.

1 Suffolk : BAKER.

The animal which is the foundation cow of this tribe shows a record, on old Suffolk dairy stock, of three generations of sires mentioned in the pedigrees of Doncaster 50, Perfection 140, and King Lud 97. The last-named bull was bred in 1873 by the Mr. Baker from whom the foundress of this tribe has since been purchased.

2 Suffolk : BOON.

The Red Polled Cattle which make up this tribe were carefully bred and selected by one man, during a period of more than twenty-five years, but the record of the bulls used in succession is imperfect.

The Editor begs to acknowledge the kindness of J. J. Colman, Esq., M.P., in permitting the use of the photograph of his cup winning heifer Dolly, impressions of which, by Messrs. Mann and Adcock, form the frontispiece of Part 2 of this Volume.

DAVYSON 3RD, 48.   SILENT LADY (O 9), Reg. No. 1855.   DOLLY (N 2), Reg. No. 1463.

A GROUP OF RED POLLED CATTLE.

RED POLLED CATTLE
FOR THE STALL AND THE DAIRY.

At the request of the Editor of the "LIVE STOCK JOURNAL" I prepared an article on Red Polled Cattle for the "LIVE STOCK JOURNAL ALMANAC" of 1883. That Almanac being now out of print, Messrs. Cassell and Company (Limited) have kindly consented to my re-issue of the article, to which I have made important additions, and have generously allowed me the use of the electro block of the Group of Red Polled Cattle, which was the frontispiece of the Almanac.

HENRY F. EUREN,

June, 1883. *Editor Red Polled Herd Book.*

The History of Red Polled Cattle can be carried back far into the last century. Suffolk had from time immemorial its breed of polled cattle producing butter which, 150 years ago, was asserted to be "justly esteemed the pleasantest and best in England." Arthur Young, in his "Survey" (A.D. 1794), defines the area—"a tract of country twenty miles by twelve, ... the seat of the dairies of Suffolk"—which, he said, must be peculiarly considered the head-quarters of the Suffolk Polled stock, though he found the breed spread over the whole county. In this "Survey" we get the first accurate description of the breed. Though Arthur Young makes no note of Norfolk Polled Cattle, yet advertisements of sales held in and from the year 1778 prove that dairies of such animals were numerous in the county, and that they extended from the northern boundary of the Suffolk "head-quarters" well into the centre of Norfolk.

An old Elmham tenant, who survived till 1872, recollected Red Polled Cattle on the estate so long ago as the year 1780. At Shipdham, they were greatly valued from a date certainly as early. At Necton, they were kept from a remote period. The predominant breed in Norfolk at that time (see Marshall's " Rural Economy of Norfolk "—Notes written from 1780 to 1782) was, however, "a Herefordshire breed in miniature," and "the favourite colour a blood-red, with a white or mottled face." Marshall fortunately preserves for this generation a record of the process by which the excellencies of this now-extinct old Norfolk blood-red stock have been combined with the proverbial merits of the Suffolk Red Polled. He says there were several instances of the Norfolk breed being crossed with Suffolk bulls, and that the result was "increase of size and an improvement of form."

A Holkham tenant, Mr. Reeve, of Wighton—of whom Arthur Young speaks, as an agriculturist whose husbandry merited attention—co-operating with his neighbour, Mr. England, of Binham, would appear to have thought more highly of this cross than did Mr. Marshall. The result of his selection was first shown in public at the Norfolk Agricultural Society's meeting, held at Swaffham, July 16th, 1808, at a time when the rage for Devons was nearly at its height on the Holkham estate. The official report of the meeting was advertised. It spoke of the bull shown by Mr. J. Reeve as follows :—" This breed is a new kind, partaking of the best qualities of the Suffolk and Devon and the old Norfolk. It has no horns, is of a true Devon or Norfolk red, and will get stock to raise fat to about fifty or sixty stone, with as little coarse meat as can be expected." Mr. Reeve could have had no part in drafting this report, or the word Devon would not have been found there; for an old letter in my possession, written by one who well knew Mr. Reeve's likes and dislikes, says " he certainly *never used* a Devon bull," and the writer goes on to speak of Mr. Reeve's " antagonism " to that breed. This " new kind " of cattle was carefully selected and bred by Mr. Reeve till September, 1828, when his dairy numbered twenty-five head, the bull, then sold, being " one of the most perfect animals in the kingdom." An equally judicious breeder was Mr. G. B. George, of Dunston, and afterwards of Eaton, near Norwich. Some of the animals were within a few years introduced into Suffolk, for crossing with the red cows there. The mixture of the two varieties has continued to this day, so that it would now be difficult to find stock which could be said to be free from its influence. Occasionally the evidence of the old Norfolk variety is made manifest by reversion, though the instances of this are now becoming very rare. Another cross was tried some forty or fifty years ago by Mr. Moseley, of Glemham, Suffolk. He used a Scotch bull for one generation, and then reverted to the original Suffolk breed. The evidence of this experiment is yet occasionally seen in the clouded noses in the few tribes which trace back to the cows of this once famous herd. Another experiment was made with a Devon cross; but the result in the end was found to be unsatisfactory. In fact, the animals whose breeding is known to have been true during the last fifty years or more give the best results now.

COLOUR

was in the opinion of the old fanciers of Suffolk Polls a distinctive characteristic. Mr. M. Biddell, speaking in 1862, could "recollect the time when no other colour than red would be looked at in a Suffolk cow," and in this discussion on colour it was admitted that "the red cow had established the breed." Previous to that meeting of the Suffolk Agricultural Society there was a tendency being developed to get rid of the colour distinction. This may have arisen from the remembrance of the fact that "red and white, brindle, and a yellowish cream colour" had also been accepted colours, as representing good milkers. In Norfolk, as I have said, red was the favourite colour, but in a few districts sheeted Polls were preferred. The fashion has during the last forty years set steadily in one direction. The red, which is now recognised as the mark of excellence, is a deep, rich blood-red, and the spot of white on the udder, which Mr. George held to be a sign of good breeding, has been crossed out. The predominance of the deep red shows plainly the degree in which the old Norfolk breed has affected the Polls, and on the contrary, the freedom from horns and from white on the udder and face is evidence of the persistence of the Suffolk Polled character. The amalgamation of the two varieties—Norfolk Polled and Suffolk Polled—may with certainty be traced from the year 1846. Both counties henceforth met in an honourable competition in the show-yard. Purchase of the handsomest and truest bred red stock became the desire of all the breeders. The result of this zeal was soon made evident, not only at county shows, but also at the Royal meetings. The breed, however, continued to be without a name until the Royal Agricultural Society, at the Battersea meeting in 1862, opened classes for "Norfolk and Suffolk Polled" Cattle. This cognomen was thereupon adopted by Norfolk, but it was never accepted by the

Suffolk Society, whose practice it has been either to provide classes for "Suffolks," or—and this very recently—for "Suffolk and Norfolk Polled." This breed now having its Herd Book, and being distributed far beyond the boundaries of the two counties, is henceforth to be known as the "Red Polled," and the Register as "THE RED POLLED HERD BOOK."

THE STANDARD DESCRIPTION

of Red Polled Cattle was agreed upon by the breeders in the autumn of 1873, after my proposal to establish a Herd Book of the breed had met with ready acceptance. This Standard Description reads as follows:—

ESSENTIALS.

COLOUR.—Red. The Tip of the Tail and the Udder may be White. The Extension of the White of the Udder a few inches along the inside of the flank, or a small white spot or mark on the Under Part of the Belly by the Milk Veins, shall not be held to disqualify an Animal whose Sire and Dam form part of an Established Herd of the Breed, or answer all other Essentials of the "Standard Description."

FORM.—There should be no Horns, Slugs, or Abortive Horns.

POINTS OF A SUPERIOR ANIMAL.

COLOUR.—A deep red, with Udder of the same colour, but the Tip of the Tail may be White. Nose, not Dark or Cloudy.

FORM.—A neat Head and Throat. A full Eye. A Tuft or Crest of Hair should hang over the forehead. The Frontal Bones should begin to contract a little above the eye, and should terminate in a comparatively narrow prominence at the summit of the head.

In all other particulars the commonly accepted points of a superior animal are taken as applying to Red Polled Cattle.

THE HERD BOOK.

Registration in the Herd Book was, at the outset, easily obtainable. Personal inspection of the herds by the Editor, and his inquiry into the breeding and antecedents of the cattle, were in most instances resorted to. The result was a return of 119 bulls and 554 cows and heifers then existing in herds whose owners accepted the conditions of registry. Subsequently, in 1877, a few other herds existing before the year 1874 were also registered. When the material had been accumulated, it became a question for the Editor whether he would follow strictly the time-honoured plan begun by Coates, that of the Polled

Herd Book, or that of the Jersey Herd Book—the only registers then existing in which there was a diversity in arrangement; or whether he should strike out a path of his own which should more effectually distinguish in future generations of cattle their respective lines of descent. The more arduous task was chosen, and the result was a distinctive mode of grouping, which is found to be of the greatest value to intelligent breeders of Red Polled Cattle.

The first issue of the Herd Book was made in the spring of the year 1874, and the Volume was completed by a second issue in 1877. This Vol. I. is known as "The Foundation Volume." It contains a history of the breed, detailed notices of the Foundation Cow of each tribe, and pedigrees of 446 bulls and 1300 Cows. Its price, post free, is 15s. The first part of the Second Volume was issued in 1881. It contains the pedigrees of Bulls from 447 to 612, and of Cows from 1301 to 1966. The Second Part, completing Vol. II., is now issued. It contains the pedigrees of Bulls from 613 to 784, and of Cows from 1967 to 2613. The price of the Volume, post free, is 10s. 6d. Henceforth the Herd Book will be issued every alternate year, the first part of Vol. III. appearing in the spring of 1885.

TRIBAL DISTINCTIONS.

Each Group is distinguished by a letter of the alphabet, and further to distinguish the stock, each Tribe in a Group—that is to say, all the descendants of one cow—is marked by a number added to the group letter. Cattle bred in America, from imported stock, are distinguished by the addition of the letter A to the original group letter—thus AA, AN, &c. The only addition to the plan thus developed has been the consecutive numbering of all the cows, as is the practice in the Polled Herd Book and in the American records, answering in effect to a permanent ear-mark for every animal. In each issue of the Herd Book is shown the constitution of the several herds as then existing. Thus is seen at a glance how many animals of any one strain, or tribe, any owner has. This list of Herds and Owners is, in fact, a key to the Herd Book. Regulations have been made for the recording of new tribes, should these be formed by the use on unregistered polled stock of pure-blood bulls for at least five generations. While there is thus ample freedom for those who wish to make experiments in cattle breeding, the descendants of such stock will be always distinguished from the cattle in the foundation tribes.

DIVERSITY OF TYPE.

Many of the old Suffolk Polled Cattle were much more massive beasts than the Norfolk; and this characteristic is yet in evidence. They could easily be picked out from a collection by the comparative coarseness of the head—a difference which is now but seldom manifest. In other points there were few divergences in character between the two varieties.

The Powell cattle have more especially been noteworthy for fineness of bone, shortness of leg, round barrels, good hind-quarters, and general neatness of outline; so that, though small, they have always won the favour of breeders. Through Norfolk Duke, a bull of Mr. Powell's breeding, his stock have influenced almost every herd in the two counties. The bull Davyson 3rd, illustrates the Powell type admirably.

The Eaton Cattle, which have yet many representatives in the herds, are depicted in "Youatt on Cattle" in a drawing made many years ago. It will be seen by a comparison of the two drawings that the type has been but little altered in the interval. There has been a great improvement in the spring of the rib and in the outline of the carcase, but otherwise the characteristics of the breed remain as they were. Clean, thin, short legs, a clean throat with little dewlap, a springing rib with large carcase, a large udder, loose and creased when empty, milk veins very large and rising in knotted puffs to the eye, are points in a good Red Polled cow now as they were in Arthur Young's day. The deep red harmonises well with the landscape, and the docility of the cows especially recommends them as park cattle.

As graceful as the Devon, the Red Polled has the additional advantage of hornlessness—in itself no little gain where horses also run in the pastures, or where the stock sent to market have to make a long railway journey.

WEIGHT.

At the close of the last century the animals when fattened seldom exceeded fifty stone (of 14lbs.) This is the report both of Marshall and Young. The former says, "The superior quality of their flesh, and their fatting freely at an early age, do away with every solid objection to their size and form." There has been great improvement in this matter of weight for age, while there has been no deterioration in the quality of the flesh: butchers now, as then, purchase the Red Polled readily, because they die well, and the meat is equal to the best

Polled Scot or Highlander. A few of the recorded weights of fat beasts will show this:—

Mr. A. Taylor's Red Polled steer, first prize at the Smithfield Club Show of 1881 (aged 3 years 7 months: sire Norfolk; dam Suffolk), had a recorded live weight of 17 cwt. 1 qr. 1 lb. [1919 lbs]. Its dead weight was 91 st. 6 lbs. (14 lbs. to the stone): a percentage of 66·74 of the live weight. The same exhibitor's heifer (aged 3 years 1 month 3 weeks) had a live weight of 13 cwt. 3 qrs. 14 lbs. [1434 lbs]. Its dead weight was 72 st. 7 lbs: a percentage of 65·31 of the live weight.

Mr. J. J. Colman's prize cow Fanny (aged 10 years 3½ months), which had produced five calves, had a live weight of 17 cwt. 22 lbs. [1926 lbs.], and was sold by public auction at Ipswich at a sum which equalled 4·375d. per pound, calculated on the live weight.

The dead weight of a 3 years 9 months old Norfolk steer, bred by Mr. T. Brown, and shown at Norwich in 1878 by H.R.H. the Prince of Wales, was 80 st. 4 lbs,; of Mr. A. Taylor's 3 years 10 months old steer, first prize winner at the same show, 111 st. 12 lbs.

This record is nearly equalled by that of a bull of Mr. Lofft's breeding, which when slaughtered in "fair condition only" gave a dead weight of 110 stone. A 2 years 9 months old steer has given a dead weight of 68 st. A 2 years 3 months old heifer a dead weight of 60 st.

The live weight of a three years old steer of the Biddell strain, shown at the Suffolk Club Show of 1876, was 25 cwt. 2 qrs. [2856 lbs.]; girth nearly 9 feet. Steers of the Davy tribe, shown at Norwich, November, 1882, had the following recorded live weight—age 2 years 11 months, 14 cwt. 3 qrs. 8 lbs. [1660 lbs.]; age 4 years 1 month, 16 cwt. 4 lbs. [1796 lbs.]; and a heifer, Davy 26th, age 3 years 7 months, 15 cwt. 2 qrs. 23 lbs. [1759 lbs].

A steer of the Elmham strain, age 3 years 10 months, which, at the same Show, won the Cup as the best Red Polled, and also the Cup as best Beast of any Breed bred and fed in Norfolk, gave a live weight of 17 cwt. 2 qrs. 18 lbs. [1978 lbs.] At the Birmingham Show, in competition with Sussex steers, including the steer which, eight days later, won the Sussex breed cup at Islington, this animal won first place; and subsequently the breed cup at the Smithfield Club Show.

Steers of the Starston tribes, at the Norwich Show, had recorded a live weight, at the age of 1 year 11 months, of 10 cwt. 1 qr. 16 lbs. [1164 lbs.]; age 2 years 11 months, 14 cwt. 3 qrs. 10 lbs. [1662 lbs.]; and a heifer, age 2 years 1 month 2 weeks, 11 cwt. 3 qrs. 6lbs. [1322 lbs].

A 3 years 6 months old steer, of the Thursford strain, had a live weight of 16 cwt. 23 lbs. [1815 lbs.]

Olivia, a heifer of the Hammond selection, which is believed to combine a strain of Devon with Red Polled blood, had a live weight, at 1 year 9 months 6 days, of 10 cwt. 3 qrs. 10 lbs. [1214 lbs].

A 4 years 1 month 1 week old cow, sired by Davyson 3rd—Blossom, 1327, had a live weight record of 16 cwt. 1 qr. 2 lbs. [1822 lbs]. Another cow from Troston, also sired by Davyson 3rd, weighed 12 cwt. 2 qrs. 23 lbs. [1423 lbs.], age 4 years 4 months.

An 18 months old Red Polled steer, bred by Mr. Robt. Savage, Bixley, and sired by May Duke 348, had a dead weight of 53 sts. 13 lbs. (14 lbs. to the stone), and its internal fat was extraordinary for an animal of that age, and not fed for show.

The percentage of dead to live weight has ranged from 65 to 66·75 for all animals of the Red Polled breed whose dead weight has been compared with live weight.

Cross-bred steers, exhibited at Norwich in November, 1882—the get of a Shorthorn bull out of Red Polled unregistered cows—had a live weight respectively of 17 cwt. 3 qrs. 20 lbs. [2028 lbs.] at 2 years 7 months, and 20 cwt. 1 qr. 2 lbs. [2270 lbs.] at 3 years 5 months.

WEIGHING LIVE CATTLE.

English farmers are very seldom provided with a machine on which to weigh live cattle, and the weight record of Red Polled Stock in store condition is thus very limited. Mr. Tyssen Amherst, M.P., of Didlington Hall, having such a machine, his farm steward, Mr. John Wallis, has at my request weighed and measured several cattle in the Didlington herd with the following results, the stock living entirely on the grass of very poor land :—

Name.	Age.	Weight.	Length from point of Shoulder.	Total Length.	Girth.
BULL.	yrs.	lbs.	ft. in.	ft. in.	ft. in.
Davyson 3rd	9	2093	5 2	7 10	7 10
COWS.					
Davy 24th (H 1)	5	1344	4 9	6 9	6 9
Dolly (P 9)	5½	1320	4 6	—	6 4
Wild Briar (B 9)	6	1436	4 11	—	6 8
Pretty Flower (B 18)	6	1427	5 0	—	6 7
Pansie (B 20)	3	1281	—	—	—
Bertha (A 20)	3	1354	—	—	—
Cheerful (K 19)	7	1514	5 0	—	6 8
Nancy 2nd (K 19)	8	1650	5 2	—	6 6
Countess (L 11)	3	1350	—	—	—
Dolly (N 6)	6	1472	—	—	—
Nancy (N 15)	9	1649	—	—	—
Satin (T 7)	3½	1358	4 8	6 7	6 9
Norfolk Witch (W 14)	5	1387	4 7	—	6 7
Poppy (U 3)	2¾	1484	4 11	6 10	7 1

Slasher 577, bred by Mr. Lofft—combining Norfolk and Suffolk blood—had a live weight of 27 cwt. [3024 lbs.] at the age of 4 years 7 months ; girth, 8ft. 2 in. His son, Rollick 558, of the same tribe as Dolly—N 2 (see portrait), weighed, at the age of 2 years 8 months 18 weeks, 19 cwt. 3 qrs. 14 lbs. [2226 lbs.], and its dead weight was 100 stones of 14 lbs. The bull, Cortes 645, weighed, when 1 year 8 months old, 12 cwt. 20 lbs. [1364 lbs.] ; eight weeks after, his live weight was 12 cwt. 3 qrs. 9 lbs. [1437 lbs.]: girth, 6ft. 10 in. King Egbert 688, at 15 months 3 weeks, weighed 10 cwt. 3 qrs. 2 lbs. [1206 lbs.] ; girth, 6 ft. 6 in. Three bull calves at Didlington, under five months old—all the get of Davyson 3rd—had a live weight of 3 cwt. 1 qr. [364 lbs.], 3 cwt. 14 lbs. [350 lbs.], and 3 cwt. 12 lbs. [348 lbs.] respectively.

A Davy heifer at Didlington, sired by Davyson 7th, and calved January 27th, 1882, had on May 31st, 1883, a live weight of 8 cwt. 1 qr. 14 lbs. [938 lbs]. ; girth, 6 ft. 1 in. A Primula heifer, calved January 3rd, 1883, weighed, on the following May 31st, 3 cwt. 1 qr. 20lbs. [380 lbs]. A Red Polled calf at birth has been found to weigh 3 qrs. 10 lbs. [94 lbs.]

MILK AND CREAM TESTS.

The testing of cattle for milk and cream is not systematically carried out by English farmers. Mr. H. Biddell has taken periodical tests during a long period, but has not made the results public, his object being rather to form an accurate personal opinion of the cows best worth retaining for the future increase of the herd. It needs no argument to prove that a cow which yields an average of 25 to 30 pints of milk per day, during eight months after calving, pays better than one which gives a good pail of milk during the first three months, and then almost dries up; or that a cow which has an average of 10 pints is much less valuable, either for breeding purposes, or as an investment, than one whose average for the same time is 25 pints. The measure or weight of milk should consequently be taken daily, or at least at weekly intervals, of every cow, as soon as she has produced her second calf, if the improvement of milk secretion be one of the things kept in view by breeders of Red Polled Cattle.

The following returns have been made in response to a request to be furnished with facts as to the yield of milk. The first record is the yield, day by day, of Davy 27th, after giving birth to her second calf. I personally selected her for experiment, in July 1882, to test the value of Guenon's theory of the escutcheon as the outward sign of a good milking strain of cattle. With, at that time, a general knowledge only of the escutcheon—so fully illustrated in the translation of Guenon's latest statement of his theory, made by Mr. T. J. Hand, Secretary of the American Jersey Cattle Club—I remarked to Mr. Jno. Hammond, the owner of the herd, that this cow, Davy 27th, should be a good milk producer, if the theory was worth anything for Red Polled Cattle. He very kindly consented to make a prolonged test for me, with the result seen below. The cow had the ordinary feed of all the stock in the herd—was out at grass (good) during the summer and autumn months, and in the winter had roots, cut straw and hay, and a little cake. The return, and that made by Mr. Wallis, from the Didlington herd, illustrate one of the most noteworthy characteristics of the Red Polled breed—namely, that most of the cows regularly yield a good quantity of milk from the birth of one calf to that of another, rather than a large supply for a short period only. My thanks are due to Mr. Hammond and to Mr. Wallis for personally interesting themselves to make the record trustworthy and complete.

RED POLLED CATTLE

DAVY 27TH — H1.

Register Number 1451.

DAILY YIELD OF MILK IN PINTS.

DAY OF MONTH.	AUGUST.	SEPTEMBER.	OCTOBER.	NOVEMBER.	DECEMBER.	JANUARY.	FEBRUARY.	MARCH.	APRIL.
1st		48	46	41	40	39	86	84	26
2nd		48	46	41	40	89	86	84	24
3rd		48	42	41	40	89	86	84	24
4th		48	42	41	40	89	36	84	24
5th		48	40	41	40	39	36	84	24
6th		48	40	41	89	39	36	84	23
7th		48	38	41	89	39	36	84	23
8th		52	38	41	89	39	36	84	23
9th		56	88	41	89	39	36	84	23
10th		56	38	41	89	39	36	84	23
11th		56	88	41	89	39	36	84	23
12th		56	88	41	89	89	36	33	23
13th		56	38	41	89	89	36	33	23
14th		52	37	40	89	89	35	33	23
15th		46	37	40	39	39	35	33	23
16th	Calved	44	37	40	39	39	35	33	23
17th		44	36	40	39	39	35	33	23
18th		42	36	40	89	89	85	33	22
19th		40	35	40	39	89	35	33	22
20th	16	40	85	40	89	89	31	33	22
21st	16	40	85	40	89	89	34	33	22
22nd	20	44	35	40	89	89	34	33	22
23rd	34	48	85	40	89	39	34	33	22
24th	42	48	35	40	89	89	34	33	22
25th	42	48	32	40	89	38	84	33	22
26th	42	48	35	40	39	38	34	33	22
27th	48	48	35	40	89	37	34	33	22
28th	44	48	86	40	39	36	84	33	22
29th	44	48	86	40	39	36	...	26	22
30th	48	46	86	40	89	36	...	26	22
31st	48	...	86	...	39	36	...	26	...
Daily Average for Month.		49·93	37·45	40·43	39·1	38·5	85·0	32·6	22·8

Daily Average for Five Months . . . . 41·04 pints.
,, ,, Six ,, ... 40·1 ,,
,, ,, Seven ,, ... 39·01 ,,
Total yield from September 1st to March 31st inclusive, 8,273 Imperial Pints,
to April 30th, 8,957 Imperial Pints.

The following returns are from the Didlington Herd (Home Farm), for which I am indebted to Mr. John C. Wallis, who keeps a careful milk and butter record.

AVERAGE DAILY YIELD OF MILK IN PINTS.

NAME OF COW.	DATE OF CALVING.	SEPTEMBER.	OCTOBER.	NOVEMBER.	DECEMBER.	JANUARY.	FEBRUARY.	MARCH.	APRIL	TO MAY 21ST.
Wild Rose Cousin	(4th Calf) Aug. 28th	40	38¼	38	36	34½	34	32	26	16
Golden Locks	(2nd Calf) Sept. 7th	42	41	40	39¼	37	36½	35	32	26
Gentle Rose	(3rd Calf) Dec. 17th	...	...	...	34	32½	32	31½	30¼	28
Pansie	(3rd Calf) Jan. 4th	...	...	...	...	38	36	34½	33¼	32

Nancy 2nd (K 19) dropped her fourth calf on August 9th, 1881. In the week ending February 5th, 1882, she gave 210 pints of milk; percentage of cream, as indicated in a graduated test-tube after the milk had been at rest 24 hours, 16·5. Each of the cows in the herd had in February a daily feed of 4 lbs. mixed linseed and decorticated cotton-seed cake, 4 lbs. bran, 1 bushel carrots, and 1½ bushels barley straw and hay chaff. This cow, Nancy 2nd, when in full profit, August 31st, was giving 36 pints of milk per day.

Davy 24th (H 1), shown three years in succession, dropped her second calf on January 27th, 1882, and gave a daily average yield of milk from that date to August 31st of 42 pints; percentage of cream, 18.

Cherry Leaf (V 3) dropped her third calf on May 16th, gave to August 31st an average daily yield of 42 pints of milk.

Flirt 3rd (V 1), a cow of similar breeding to Cherry Leaf, gave, six weeks after producing her first calf, a yield of 249 pints of milk in the week; percentage of cream, 15.

Waxwork 6th (U 9) (the tribe in which the bull Slasher above-named is included) produced her first calf on January 8th, and on August 31st was giving milk which yielded 21 per cent. of cream.

The following returns are from the Necton Hall herd (Mr. R. H. Mason's). In the third week of February the cows were on pasture (very light land) most of the day, with a few roots; at night they each received 7lbs. cotton cake and spiced cake, 7lbs. bran, 14lbs. hay and cut straw.

Nancy 3rd (N 3), aged six years, dropped her calf in December, 1881; on

February 18th yielded 28 pints of milk at two successive milkings: percentage of cream, 16.

Pet (N 1), aged six years, dropped her calf January 22nd; on 18th February yielded 23 pints of milk; percentage of cream, 35.

Tulip (N 4), with similar conditions, yielded 25 pints of milk; percentage of cream, 34. And Tulip (N 7), aged 9 years, which dropped her calf in October, 1881, was yielding 26 pints of milk per day in February.

Tests were also taken at the end of August, when the cows were all at grass, with the following results:—

Empress (N 4), which dropped her third calf on April 10th, yielded 22 pints of milk per day; percentage of cream, 29. Sultana (N 5), which dropped her fourth calf on March 22nd, gave 30 pints; percentage of cream, 26. The butter being produced by eleven cows in August was 80 lbs., and 120 pints of new milk were sold per week. In the year 1881, from the herd of thirteen Red Polled cows, eight heifers, and one Alderney, the produce of marketable butter was 3,120½ lbs.; new milk sold, 725 gallons; cream sold, 101 pints; money value, independent of skim milk, £260. In the year 1882 from fourteen cows, six heifers, and one Alderney, the produce of marketable butter was 3,434 lbs.; new milk 686 gallons; cream 13¼ gallons. The money realised was £281. 4s. 2d.

Primrose (K 6), an eleven-year-old cow in Lord Kimberley's herd, gave on winter feed (hay, chaff, bran, and cake), six weeks after calving, 32 pints of milk per day, and the marketable butter produced was 9lbs.

Mr. Lofft, Troston Hall, reported the testing of two of his cows of the Handsome (U 3) tribe, each of which consumed per day 4 lbs. cotton cake, 2lbs. Simpson's meal, 6 stone of beet root, and 1½ bushels of chaff. Handsome 5th, four months after calving, yielded 28 pints of milk per day and 7lbs. of marketable butter per week. Handsome 6th yielded 32 pints of milk per day and 10 lbs. of butter per week.

Mr. G. Gooderham, Monewden—Cherry Leaf (V 3) and Flirt 3rd (V 1), were bred in his herd—uniformly causes his cows to breed very early, and the secretion of milk is thus fostered. One of his cows, Wild Rose of Kilburn (V 1), which was first prize winner as a yearling at the Royal meeting of 1879, produced her first calf when wanting two days of being two years old. Before she was three years old she produced a second calf, and again within twelve months a third. Eight weeks after this last calf was dropped she gave 30 pints of milk per day on winter feed, and her average of butter was 9lbs. per week, taking all the year, since she never goes dry. In June, 1882, six months after calving, she won first prize at the Essex show as a milker; her dam won a like honour at the Suffolk show in June, 1881.

The herd of Mr. J. J. Colman, M.P., at Easton Lodge Farm, near Norwich, which has seven times in eight years won the cup offered at the Norfolk Show for the best collection, includes the seven-year-old cow (Silent Lass—O 9, dam of Silent Lady), the yearling heifer shown in our illustration. This cow, on winter feed, gave 37 pints of milk per day, eight weeks after calving. In May, when the cows were at grass—(very poor herbage, growing on a marsh)—I tested the quality of the milk, using for the purpose Heeren's Milk Tester, the "Pioskop" of the Hanover Vulcanite Company. The milk was drawn on to the Pioskop direct from the udder, when milking had been half done. Silent Lass, five months after calving, yielded milk which contained more fatty particles than are found in rich milk as marked on the tester. Even the first milk drawn from the udder of Dolly, six months after calving, was "normal" according to the tester, and her average yield was very rich, as was also the yield of the other cows tested—Rosa (P 3), seven months after calving, and Rosebud 2nd (K 17), nine months after calving.

Mr. Garrett Taylor's large herd at Whitlingham, near Norwich, is kept exclusively for the supply of milk to customers in the city. The Cafes, which have a large demand for the article, have familiarised the public with the fact that the milk of the Red Polled Cattle is exceptionally rich. One of the Whitlingham cows, on winter feed, five weeks after calving, gave 32 pints of milk per day; another 27 pints.

Mr. B. Stimpson, of Morton, reported two of his cows, on winter feed—Cheerful (W 3)—as yielding daily, ten weeks after calving, 30 pints of milk; Silky (F 4), six weeks after calving, 26 pints; and the butter made from their milk in the week 14½lbs. A four-year-old cow of the Eaton Strain, in Mr. J. F. Rogers' herd, at Swanington, yielded, five weeks after calving, on very poor food—hay, pulped swedes, and cut straw, with 3lbs. of decorticated cotton cake—28 pints of milk per day. His herd of seven cows (six Red Polled and one Shorthorn) produced in the year ended 30th September, 1882, 1,435lbs of butter, which, with milk sold amounting to £11. 18s. 10d., made the total return £118. 15s. 3d. A return of the test of two cows of the Glemham strain (Mr. Moseley's) already mentioned (in Mr. J. M. Spink's herd, Harpley) gave 53 pints of milk as the yield per day on winter feed, and 23lbs. 2ozs. of butter per week.

PREPOTENCY OF THE POLLED TYPE.

Red Polled Cattle are found to lay on flesh rapidly on pasture of the poorest character, where other breeds need to have an additional supply of richer food. The dry temperature of Norfolk and the poor pasture seem more particularly to have had their effect on the size of the stock. The first cross—stock sired by a Red Polled bull, no matter of what horned breed is the dam, is usually red in colour and polled in character. Such animals, when fat, are eagerly bought by the butcher. I have recently seen a number of such cross-breds, the produce of a Red Polled bull and a pure-bred Jersey cow, and am told the cross is an excellent one. Some of the animals had a few silver hairs mixed with the red coat; all were polled, and all had black noses. General L. F. Ross, formerly of Illinois, now of Iowa City, Iowa, has reported the results of using a Red Polled bull on Devon cows to be equally satisfactory—the cattle are all blood-red and polled.

In the year 1861, Sir E. C. Kerrison sent Lord Dartrey a number of Red Polled Cattle for his estate in Co. Monaghan, Ireland. Lord Waveney, in 1873, found that of eight or nine cows which had been bred in Ireland, although they had been crossed with other stock, only one failed to present all the points insisted upon as "Essentials" of the Red Polled breed.

But a more remarkable illustration of the persistency of the polled character is seen in the descendants, yet in America, of a Red Polled cow, sent to Ireland before the year 1847. The *Live Stock Journal Almanac*, for 1882, reports hornless cattle, known in the United States as "Jamestown" stock. A Suffolk heifer was presented in 1847 by the Lord Lieutenant of Ireland to Capt. R. B. Forbes, of the ship Jamestown. She proved to be a deep milker. Though she was bred for several years to Jersey and other bulls, nearly all her progeny—and with one exception they were bulls—were polled. In 1878, Mr. A. W. Cheever, of Sheldonville, Mass., had a polled herd, the get of a polled bull, which was descended from the imported Suffolk heifer. The cows were uniform in appearance, although the Jersey blood sometimes showed in the colour. The article is illustrated with an engraving from a photograph of a six-year-old cow of this mixed family, in which the polled character is seen to be strongly marked.

"Muley" cattle have been in Virginia for a great many years, and their descendants have also been uniformly polled. The use of a Red Polled bull has speedily brought the young stock to the desirable uniformity of colour. It would be of value to the students of the history of cattle were search to be made respecting the introduction of polled stock into America. It is recorded that many of the earlier settlers were natives of Norfolk and Suffolk villages. May they not have taken over the polled cattle, which in that day were so numerous in Suffolk and on the Norfolk borders?

The chief hindrance to the extension of the breed exists in

THE SCARCITY OF THE STOCK.

This has, in great measure, arisen from the fact of rinderpest having a few years ago been fatal to a large proportion of the cattle then in the more noteworthy herds. Fashion also had a marked effect. Shorthorns and Devons were at one time in such favour that Polled Cattle were despised and their merits ignored. With registration, however, and marked progress made in Red Polls within the last ten years, the shortness of numbers is being in some measure compensated for, noblemen and gentlemen now sparing no pains to make the breed a success.

THE ILLUSTRATION.

Davyson 3rd (48), the bull shown in the illustration, was bred by Mr. John Hammond, of Bale, East Dereham; was sold as a two-year-old to Mr. J. Foster Palmer, and on his decease was bought at auction by Mr. W. A. Tyssen Amherst, M.P., at 205 guineas. He was calved in August, 1873, being of the Davy (H 1) tribe, and got by a bull of Powell blood, sire and dam. He was the reserve at the Norfolk Show of 1875, and since that year has never been beaten at a Royal or County Show, winning sixteen first prizes and six cups. He has been, and is yet, a good stock getter, and his progeny have been most successful in the show-yard.

Dolly (N 2), calved November 3rd, 1879, the older of the two females in the illustration, was in Mr. Colman's cup collection in 1881, and again in 1882. In each year she was first in her class, and in June, 1882, she also won the cup offered for the best Red Polled Cow or Heifer at the Norfolk Show. She is a heavy-fleshed animal, inheriting that characteristic from her great-great-grand-dam, Minnie, the foundress of a Necton tribe, and herself the daughter of the Red Polled bull which won first prize at the Norwich Royal in 1849. This Minnie tribe realises high prices, and is as a rule very good both for milk and for beef. The sire of Dolly, and also of the other female in the illustration, was Rufus, a bull of Powell's famous Rose tribe, bred by the late Lord Sondes, from a grand cow now 14 years old, and owned by Mr. Fulcher, Elmham. Dolly produced her first calf in November, 1882.

Silent Lady (O 9), calved December 18th, 1880, the yearling heifer shown in the illustration, was also in Mr. Colman's cup collection of 1882. She traces back to one of Sir E. C. Kerrison's grand old-fashioned cows—a superior milker. To this old blood has been added two successive mixtures of Powell blood, with the happiest effect.

The bull has been drawn from a photograph taken from life by T. Smith, of King's Lynn, in March, 1881, when the animal was in store condition; the females from photographs from life taken in June, 1882, by Mann and Adcock, of Norwich.

A STUD BOOK

For ROADSTERS, HACKNEYS, COBS, AND PONIES.

First Issue in 1884. *Editor*, H. F. EUREN.

Every FRIDAY, Price 4d.

THE LIVE STOCK JOURNAL
AND
FANCIER'S GAZETTE,

A Chronicle for Country Gentlemen, Breeders, and Exhibitors.

The LIVE STOCK JOURNAL is the official organ of the ROYAL AGRICULTURAL SOCIETY, the SHORTHORN CLUB, the SMITHFIELD CLUB, the BRITISH DAIRY FARMERS' ASSOCIATION, the ENGLISH CART HORSE SOCIETY, the CLYDESDALE SOCIETY, and other Associations, and now stands without a rival as a Chronicle for Country Gentlemen, Breeders, and Exhibitors.

Amongst the subjects which are fully treated in THE LIVE STOCK JOURNAL *by Eminent Practical Writers are—*

HORSES,	RABBITS,
CATTLE,	PETS,
SHEEP,	DAIRY FARMING,
PIGS,	POULTRY FARMING,
DOGS,	ARTIFICIAL INCUBATION,
POULTRY,	PASTURES, The Utilisation of,
PIGEONS,	FISH CULTURE,
AVIARY,	NATURAL HISTORY,
GOATS,	&c. &c. &c.

Terms of Subscription by Post—Three Months, 4s. 11d.; *Six Months,* 9s. 9d.
Twelve Months, 19s. 6d.

Publishing and Advertisement Offices:
LA BELLE SAUVAGE YARD, LUDGATE HILL, E.C.

HERDS OF REGISTERED RED POLLED CATTLE.

Registered Cattle, existing in Herds,
Arranged Alphabetically, in Order of Tribes,
WITH THE
Registered Number of each Animal.

The **Group Letter and Number**, and the **Tribal Name**, are printed in Black Capital Letters. When one or more Bulls are included in the list of Cattle of any one tribe their Names are printed in Small Capitals, while the Names of Cows are printed in ordinary type.

REV. S. ALLEN, D.D.,

Shouldham Hall, Norfolk.   Post Town—Downham Market.

A 1 : PRIMROSE—Shouldham Primrose 5th 1181. Shouldham Primrose 6th 1182. Shouldham Primrose 7th 1183. Shouldham Primrose 8th 1184. Shouldham Primrose 9th 1185. Shouldham Primrose 10th 1844. Shouldham Primrose 11th 1845. Shouldham Primrose 12th 1846. Shouldham Primrose 13th 1847. Shouldham Primrose 15th 2529. Shouldham Primrose 16th 2530. Shouldham Primrose 17th 2531. Shouldham Primrose 18th 2532. Shouldham Primrose 19th 2533.

REV. DR. ALLEN'S HERD—*Continued*.

M 6 : STRAWBERRY—Shouldham Strawberry 10th 1186. Shouldham Strawberry 11th 1187. Shouldham Strawberry 14th 1850. Shouldham Strawberry 15th 1851. Shouldham Strawberry 16th 1852. Shouldham Strawberry 17th 2534. Shouldham Strawberry 18th 2535.

SUMMARY:—Of the Elmham Group 14 Cows. Marham Group 7 Cows. Total 21.

W. A. T. AMHERST, ESQ., M.P.,

Didlington Hall, Norfolk.   Post Town—Brandon.

Steward—Mr. JOHN C. WALLIS.

A 1 : PRIMROSE—Troston Nelly 2576.

A 2 : CHERRY—Moss-rose Queen 1685.

A 6 : NORTON—Nettle 1705.

A 11 : NANCY—KING EGBERT 688.

A 27 : CURSON—Berenice 2016. Blueberries 2029. Blueberries 2nd 2030.

B 9 : ROSE—EARL DAVY 665. Gentle Rose 914. Gentle Rosy 2209. Sweet Briar 2565. Wild Briar 1269.

B 10 : BURY—SILVERSIDES 750. SIR SAMUEL 754. Golden Locks 1545.

B 11 : SUFFOLK—Beatrice 1308. Beauty 2003.

B 18 : FANCY—Pretty Blossom 2465. Pretty Flower 1093.

B 20 : PICKET—Pansie 1717. Pattie 2429. Pattie 2nd 2430.

H 1 : DAVY—DAVYSON 3rd 48. DIDLINGTON DAVYSON 656. DIDLINGTON DAVYSON 2nd 657. Davy 24th 1448. Davy 30th 1454. Davy 31st 1455. Didlington Davy 2148.

I 2 : RUBY—Jessie Brown 1596.

I 13 : ROSEBUD—Rosy 2512.

MR. AMHERST'S HERD—*Continued*.

K 19 : ROSE—Sir David 751. Charming 2074. Chazalie 2076. Cheerful 762. Cheery 1372. Cheering 2077. Nancy 2nd 1691. Nanny 2402. Rose 484. Rose-leaf 1815. Rose-leaf 2nd 2505.

L 3 : ELMER—Eleanor 1477. Ellen 1480. Elm-leaf 1491.

L 11 : LETTON—Countess 1407.

N 2 : MINNIE—Mary 1014.

N 6 : TIT—Necton Duke 702. Dolly 852.

N 15 : NANCY—Nancy 1038.

N 19 : VICTORIA—Viola 2601. Violet 1253.

O 2 : QUEEN—Ruby 2nd 1824.

P 7 : VIOLET—Ringlet 1783.

P 9 : CHERRY—Sir Nicholas 753. Clary 1397. Dolly 1464. Dora 2152.

Q 3 : WINSOME—Winsome 1274.

T 7 : STRANGER—Satin 1837. Satinette 2526.

U 3 : HANDSOME—Cortes 645.

U 5 : PRIMULA—Primula 2468.

U 9 : WAXWORK—Count Davy 646. Waxwork 6th 1932. Waxy 2604.

U 43 : POPPET—Poppinette 2455. Poppy 2456.

V 1 : COWSLIP—Lord Davy 694. Flirt 2nd 1516. Flirt 3rd 1517. Fuchsia 2204.

V 2 : RED STOCKINGS—Bonnet Rouge 628. Red Stockings 1128. Wild Rose Cousin 1938.

V 3 : FILLPAIL—Cherry-leaf 1384. Cherry Queen 2085.

V 5 : CHERRY—Sir Simeon 755. Marjoram 1661.

W 3 : NELLY—Helene 944. Hilda 2253.

MR. AMHERST'S HERD—*Continued.*

W 14 : CLARA—Norfolk Wizard 709. Cyclamen 814. Norfolk Witch 1054.

Summary:—Of the Elmham Group 1 Bull 6 Cows. Biddell Group 3 Bulls 12 Cows. Hammond Group 3 Bulls 4 Cows. Hudson Group 2 Cows. Kimberley Group 1 Bull 10 Cows. Dereham Group 4 Cows. Necton Group 1 Bull 5 Cows. Oakley Group 1 Cow. Powell Group 1 Bull 4 Cows. Stalham Group 1 Cow. Thursford Group 2 Cows. West Suffolk Group 2 Bulls 5 Cows. East Suffolk Group 3 Bulls 8 Cows. Wolton Group 1 Bull 4 Cows. Total 84.

C. AUSTIN, Esq.,

Brandeston Hall, Suffolk.   Post Town—Wickham Market.

Steward—Mr. Kersey.

A 18 : SUITOR—Miss Mariquita 2nd 1026.

B 21 : ROSEBUD—Jumbo 684. Sweet-bloom 1221.

K 24 : FLORA—Shylock 572.

M 2 : RED ROSE—Nancy 2396.

U 45 : CONSTANCE—Constance 2nd 800. Countess 2107. Lady Jane 2293.

V 2 : RED STOCKINGS—Flageolot 1514. Flora 2nd 897. Iris 2260. Shotover 2528.

W 3 : NELLY—Miriam 2373. Queen May 2479.

1 Suff : BAKER—Dido 2150. Dutch Oven 2157.

Summary:—Of the Elmham Group 1 Cow. Biddell Group 1 Bull 1 Cow. Kimberley Group 1 Bull. Marham Group 1 Cow. West Suffolk Group 3 Cows. East Suffolk Group 4 Cows. Wolton Group 2 Cows. Suffolk Group 2 Cows. Total 16.

Mr. JOHN BAKER,

Colville House, Wisbech, Cambridge.   Post Town—Wisbech.

A 4 : RINGLET—Cameo 1357. Cassie 1363. Miss Emma 2379.

M 2 : RED ROSE—Heather 1563.

M 5 : SYBIL—Salome 1836.

R 9 : BRUNDISH—Mischievous 701. Nancy 2394.

SUMMARY:—Of the Elmham Group 3 Cows. Marham Group 2 Cows. Bungay Group 1 Bull 1 Cow. Total 7.

Mr. JOHN BALY,

Hardingham, Norfolk.   Post Town—Attleborough.

A 11 : NANCY—Fortuna 1524. Lady Legge 2295.

A 27 : CURSON—Lady Fulcher 2291. Lady Sondes 2300.

E 11 : POLLY—Brutus Duo 463.

O 1 : DUCHESS OF SUFFOLK—Brutus 4th 634. Lady Caroline 1610. Lady Kerrison 2294.

W 3 : NELLY—Brutus 3rd 633. Eleanor 1478. Lady Nelly 2298.

1 Norf. : POND—Handsome of Tittleshall 2235. Lofty 2322. Mrs. Case 2384. Troublesome 2577.

3 Norf. : NICHOLSON—Gressenhall 2222. Handsome of Gressenhall 2236. Lady Fanny Case 2290. Long Tail 2323. Mrs. Nicholson 2385. Nancy 2400.

4 Norf. : PECK—Lady Peck 2299.

SUMMARY:—Of the Elmham Group 4 Cows. Eaton Group 1 Bull. Oakley Group 1 Bull 2 Cows. Wolton Group 1 Bull 2 Cows. Norfolk Group 11 Cows. Total 22.

HERDS AND OWNERS.

MR. BANTOFT, Jun.,

Martlesham Hall, Suffolk.   Post Town—Woodbridge.

B 2 : CHERRY LUX—King Bramble 686.

MR. HERMAN BIDDELL,

Playford, Suffolk.   Post Town—Ipswich.

A 14 : TIT—Curly Coat 1423.

B 2 : CHERRY LUX—Romeo 742. Blackberry Jam 1324. Cherry Pie 787. Christmas Bloom 793. Currant Wine 1424. Early Bloom 1473. Magpie 1652. March Bloom 1653.

B 4 : WRYNECK—Little Bird 1629. Pretty Bird 2464.

B 5 : LOCKET—Bracelet 718.

B 7 : LILY—Marigold 1660. Sunflower 2560. Tiger Lily 1232. Tiger Cat 1892.

B 8 : HANDSOME—Baroness 687.

B 9 : ROSE—Abbot Sampson 613. Sweetmeat 760. China Rose 788. Gentle Hands 1536. Guelder Rose 2224. Honey Suckle 1574. Last Rose 1614. Moss Rose 2383. Old China 1713. Queen Rose 1759. Standard Rose 1867.

B 10 : BURY—Auburn Locks 1306. Light Hair 2311. Silver Star 1191.

B 11 : SUFFOLK—Bella 2009. Bellona 705.

B 12 : BEE—Honey Bee 1572. Last Bee 2302. Shining Hours 1842. Sister Anne 2539.

B 17 : FAIRY—Cinderella 1394. Coral 2102. Pallas 1716. Portia 2458.

B 18 : FANCY—Prince Charlie 730. Choice Flower 1392.

B 20 : PICKET—Blue-Beard 624. Orlando 711. Bezique 2019. Carnation 745. Picotee 410. Piquet 1076.

MR. HERMAN BIDDELL'S HERD—*Continued*.

B 21 : ROSEBUD—Bouquet 1354. Perfume 1731. Vinaigrette 2599.

R 1 : COSSET—Susanna 1219. Barmaid 1996. Jessica 2267.

V 5 : CHERRY—Timon 764. Celia 2062. Juliet 1603. Sweetheart 2566. Trimley Cherry 1240. White Heart 651. Wistful 1944.

W 9 : LADY—Grand Lady 1647. Lady Little 1612. Little Lady 1004. The Nun 2421.

W 14 : CLARA—Merry Witch 1666. Witchcraft 1946.

> SUMMARY :—Of the Elmham Group 1 Cow. Biddell Group 5 Bulls 45 Cows. Starston Group 3 Cows. East Suffolk Group 1 Bull 6 Cows. Wolton Group 6 Cows. Total 67.

H. BIRKBECK, Esq.,

Stoke Holy Cross.   Post Town—Norwich.

Bailiff—Mr. W. CHENEY.

E 11 : POLLY—Ella 1479. Elmham Georgina 1487.

E 13 : BARKER—Elizabeth 2159. Elmham Taylor 1493.

P 3 : ROSE—Bon-Bon 627. Broom 781. *Brush 2045.

P 4 : NINA—Blue Bell 2028. Blue Bonnet 2031. Nina 5th 1053.

S 2 : HAPTON CHERRY—Harriet 2243.

S 3 : DOWSON—Hulver 1578. Hyacinth 1585.

T 4 : TIT—Tiny 1895. Tipple 1896. Tit 3rd 607. Tit-lark 1901. Tit-mouse 1903. Topsy 714.

T 6 : NANCY—Bee 77. B.B. 1315. Blacking 2021.

> SUMMARY :—Of the Eaton Group 4 Cows. Powell Group 1 Bull 5 Cows. Stoke Group 3 Cows. Thursford Group 8 Cows. Total 21.

* By error marked P 2 in Register.

HERDS AND OWNERS.

MR. PETER BLOFIELD,

Quidenham, Norfolk.   Post Town—Attleborough.

K 25 : BRIDE—Peppercorn 716. Patience 1723. Peony 2438. Phœbe 2441. Polly 1740. Prim 1746. Primrose 1748. Primula 1751. Prune 1116.

K 26 : FULLER—Radical 733. Rhoda 2490. Rosette 1162. Rose Princess 2509. Ruth 2522.

Summary :—Of the Kimberley Group 2 Bulls 12 Cows. Total 14.

MR. JOSEPH BOGGIS, Jun.,

Geldeston, Norfolk.   Post Town—Beccles.

R 8 : BEAUTY—Fancy 2183.
R 11 : PRETTY—Strawberry 2554.

Summary :—Of the Starston Group 2 Cows.

MR. E. BOON'S EXECUTORS,

Brandeston, Suffolk.   Post Town—Wickham Market.

U 45 : CONSTANCE—Hunston Duke 3rd 677.
2 Suff. : BOON—Lucas 696. Beauty 2007. Cherry 2nd 2081. Cossett 2nd 2104. Handsome 2237. Hester 2252. Lovely 2330. Nancy 2nd 2401. Nelly 2407. Sprightly 2548. Strawberry 2555.

Summary :—Of the West Suffolk Group 1 Bull. Suffolk Group 1 Bull 10 Cows. Total 12.

MR. WM. BRADFIELD,

Ramsley Farm, Elmham, Norfolk.   Post Town—Elmham.

A 1 : PRIMROSE—Nellie 1702.
A 6 : NORTON—Tom 766. Nettie 1046.

MR. BRADFIELD'S HERD—*Continued.*

A 33 : ELM-LEAF—Elm-leaf 1488.

N 6 : TIT—Tommy 588. Cherry 94.

1 Norf. : POND—Lancer 689. Charmer 2073. Lettice 2309. Wiffin Cherry 2606.

SUMMARY :—Of the Elmham Group 1 Bull 3 Cows. Necton Group 1 Bull 1 Cow. Norfolk Group 1 Bull 3 Cows. Total 10.

MR. T. BROWN,

Marham Hall, Norfolk.   Post Town—Downham Market.

A 1 : PRIMROSE—Maggie 329. Marian 1012. Mercy 2366. Mildred 1017. Milicent 1669. Mistletoe 2382. Myrtle 2387.

A 4 : RINGLET—Cactus 2054. Camilla 1358. Camlet 1859. Camomile 1360. Chantress 2065. Charity 2066. Cicely 2090. Cinnamon 2091. Columbine 2097. Kathleen 972. Katinska 973. Katrine 974. Keepsake 1606.

E 2 : CHERRY—Lavinia 1618. Lois 1635. Lucretia 2336. Raisin 2481. Rhoda 1778. Rosamond 1136. Rupee 2520. Theresa 1228. Truth 2578.

I 1 : BEAUTY—Ella 870. Era 2167. Erica 2168. Evelyn 1501.

I 2 : RUBY—Julian 682. Jael 1593. Jane 957. Jessamine 1594. Jessica 1595. Jet 2271. Jonquil 2275. Josephine 2276. Joyce 2277. Juanita 1600. Julia 2280. Juliet 967. Juno 2284.

M 1 : ALMA—Abbess 1967. Acacia 1971. Actress 1977. Adelaide 1978. Agar 670. Agatha 1303. Alice 4. Alicia 1304. Auburn 1991.

M 2 : RED ROSE—Hairbell 2226. Handmaid 2227. Harp 2242. Hawthorn 1561. Heath 2245. Helen 265. Heliotrope 1566. Hemp 2248. Hermia 2249. Hermione 2250. Hilda 948. Honeydew 2254. Honor 1576. Honoria 1577. Hortensia 953. Hostess 2257. Hypatia 2258.

MR. BROWN'S HERD—*Continued.*

M 5 : SYBIL—Sarah 1176.

N 2 : MINNIE—Bayard 624. Beatrice 23. Berenice 2017. Bertha 706. Blanch 2022. Elmham Belle 202.

P 1 : HANDSOME—Plato 721. Pliny 724. Nelly 372. Pastille 2427. Pauline 1726. Penelope 1069. Pensive 1728.

P 3 : ROSE—Fabian 667. Frank 493. Fusilier 672. Faith 1504. Fatima 2186. Flame 2191. Fragrance 2197. Fusee 1530.

Summary :—Of the Elmham Group 20 Cows. Eaton Group 9 Cows. Hudson Group 1 Bull 16 Cows. Marham Group 27 Cows. Necton Group 1 Bull 5 Cows. Powell Group 5 Bulls 10 Cows. Total 94.

MR. BURTON,

Lowestoft, Suffolk.   Post Town—Lowestoft.

R 10 : CHERRY—Cherry 2nd 2079.

The Most Honourable the MARQUIS of BRISTOL.

Ickworth Park, Suffolk.   Post Town—Bury St. Edmunds.

B 2 : CHERRY LUX—Cherry Bloom 2083. Cherry Bud 781.

B 3 : NANCY—Briony 60. Duchess Briony 1st 1469. Duchess Briony 2nd 1958. Princess Briony 1112.

B 6 : SWEET—Meadow Sweet 2361. Sweetheart 1883.

B 9 : ROSE—Rosa 1785. Rosabelle 2494. Rosamond 1788. Rosary 1140. Rosebud 2501. Rosette 508.

B 11 : SUFFOLK—Suffolk Countess 2556. Suffolk Princess 1875.

B 18 : FANCY—Fancy Flower 219.

B 21 : ROSEBUD—Blossom 2025. Full Bloom 1529.

B 22 : STRAWBERRY—Strawberry 1871.

HERDS AND OWNERS.

The Most Honourable the MARQUIS of BRISTOL'S HERD
Continued.

U 9 : WAXWORK—Saxham Prince 748.

V 1 : COWSLIP—Duchess Lovely 1471. Lady Love 1613. Lovable 1637. Lovely 324. Lovely Duchess 2331. Lovely Lass 1638. Lovely Queen 2332. Lovely Rose 1639. Princess Lovely 2470.

V 3 : CHERRY—Cherry 2nd 1378.

V 4 : HANDSOME—Countess Lucy 1413. Duchess Lucy 1472. Lucilla 1640. Lucinda 1641. Lucy 326. Lucy Rose 1647.

V 5 : CHERRY—Cherry Prince 640. Cherry Blossom 2084. Cherry Duchess 1382. Cherry Rose 1388. Double Cherry 855.

W 9 : LADY—Duchess Dot 1470. Fancy Spot 2184. Lady Dot 984. Lady Spotless 304.

Summary :—Of the Biddell Group 20 Cows. West Suffolk Group 1 Bull. East Suffolk Group 1 Bull 20 Cows. Wolton Group 4 Cows. Total 46.

MR. JAMES E. CARMALT,

Scranton, Pennsylvania, U.S.A.

AA 12 : HANDSOME—Martha 1662.

AU 43 : POPPET—Scranton Æsthete 749.

Summary :—Of the American Group 1 Bull 1 Cow.

MR. H. J. CHAMBERLIN,

Davilla, Milan Co., Texas, U.S.A.

AA 12 : HANDSOME—* Madeira 1650.

AU 5 : PRIMULA—Roger 738.

Summary :—Of the American Group 1 Bull 1 Cow.

* In the Register the Tribal Number of Madeira should read AA 12.

MESSRS. J. H. & W. W. CLARK,

Toledo, Washington Co., Pennsylvania, U.S.A.

AU 5 : PRIMULA—Alfred 616.
AW 14 : CLARA—Alforata 1980.

  Summary :—Of the American Group 1 Bull 1 Cow.

MR. CLARKE,

Aldeby Suffolk.   Post Town—Beccles.

R 9 : BRUNDISH—Aldeby 614.

MRS. COLLYER,

Hackford Hall, Norfolk.   Post Town—Reepham.

Agent—Mr. D'Arcy Collyer.

A 24 : FLOSS—Beatrice 2001. Nina 2411.

  Summary :—Of the Elmham Group 2 Cows.

J. J. COLMAN, Esq., M.P.,

Carrow House, Norwich.

Steward—Mr. Garrett Taylor, Trowse House, Norwich.

Bailiff at Easton Hall Farm—Mr. Hebgin.

A 9 : FANNY—Francillo 669. Roundhead 564.
B 2 : CHERRY LUX—Shylock 572.
H 4 : OLIVE—Olivia 1715.
K 17 : CHERRY—Don Carlos 659. King Charming 687. Othello
       713. Cherry Leaf 1383. Cherry Ripe 2086. Miss Atkins
       1023. Rosalie 2495. Rosebud 2nd 1797.

HERDS AND OWNERS.

MR. COLMAN'S HERD—*Continued.*

K 19 : ROSE—King Charles 329.

N 2 : MINNIE—Cromwell 647. Dolly 1463.

O 9 : SILENCE—Silence 1853. Silent Beauty 2536. Silent Lady 1855. Silent Lass 1189. Silent Woman 2537.

P 1 : HANDSOME—Hetty 1569.

P 2 : STRAWBERRY—Brown 1343. * Brunette 2044.

P 3 : ROSE—Goshawk 497. Rosa 1133. Rosemary 2508. Rosamond 1789. Rosy Morn 2514.

T 1 : PRIMROSE—Prudish 2474.

W 2 : BEAUTY—Druid 662.

Summary—Of the Elmham Group 2 Bulls. Biddell Group 1 Bull. Hammond Group 1 Cow. Kimberley Group 4 Bulls 5 Cows. Necton Group 1 Bull 1 Cow. Oakley Group 5 Cows. Powell Group 1 Bull 7 Cows. Thursford Group 1 Cow. Wolton Group 1 Bull. Total 30.

* By error marked P 3 in Register.

MR. C. K. CORDY,

Trimley St. Mary, Suffolk.   Post Town—Ipswich.

A 1 : PRIMROSE—Trimley Gem 2574.

A 21 : ROSE—Cowslip 2109.

E 3 : COUNTESS—Esther 874.

O 8 : MARY GREY—Nonsuch Tom 708. Trimley Nonsuch 1912.

R 10 : CHERRY—Nell 1700.

U 9 : WAXWORK—Not Doubtful 710. Trimley Pretty 1913.

U 43 : POPPET—Trimley Beauty 1906.

X 1 : LOVELY—Rose of Trimley 1796. Rose 2nd of Trimley 2500.

HERDS AND OWNERS.

MR. CORDY'S HERD—*Continued.*

X 4 : BEAUTY—Trimley Daisy 2nd 1908. Trimley Daisy 3rd 1909.

X 5 : BLUE-BELL—Trimley Handsome 2nd 2575.

SUMMARY:—Of the Elmham Group 2 Cows. Eaton Group 1 Cow. Oakley Group 1 Bull 1 Cow. Starston Group 1 Cow. West Suffolk Group 1 Bull 2 Cows. Trimley Group 5 Cows. Total 14.

MR. S. B. DOUGLAS,

Reed's Corner, Ontario Co., New York, U.S.A.

AA 29 : BELLE—JOSEPHUS 679.

MR. W. B. EASTER,

Stockton Hall, Norfolk.   Post Town—Bungay.

E 11 : POLLY—Lily 1625. Primrose 1099. Sprightly 2547.

I 9 : BRIDESMAID—TROSTON PRINCE 771.

R 8 : BEAUTY—Beauty 696. Fancy 1507.

R 9 : BRUNDISH—Brundish 2nd 1344. Violet 1927.

R 11 : PRETTY—Fillpail 1513. Nancy 2395. Pretty 1092. Strawberry 1873.

SUMMARY:—Of the Eaton Group 3 Cows. Hudson Group 1 Bull. Starston Group 8 Cows. Total 12.

MR. E. W. ENGLISH,

Saranac, Michigan, U.S.A.

AA 29 : BELLE—RED DUKE 554.

AU 5 : PRIMULA—Cecilia 2nd 2060.

SUMMARY:—Of the American Group 1 Bull 1 Cow.

MR. T. FULCHER,

Elmham, Norfolk.   Post Town—East Dereham.

A 2 : CHERRY—Elmham Cherry 1486.

A 11 : NANCY—Falstaff 303. Fury 495. Fame 2181. Fatima 1509. Spotless 1197.

A 18 : SUITOR—Sally Senter 2524.

A 26 : GATELY—Villebois 775. Vanessa 2590. Venus of Guist 2592. Vestal 2595.

A 27 : CURSON—Bowler 629. Becky Sharpe 2008. Bertha 1320. Brilliant 728. Brindy 729.

A 37 : FENN—Nobby 706. Ruler of Elmham 744. Nancy 1689. Ruby 1823.

B 12 : BEE—Busy Bee 2048.

E 5 : ROSE—Rosy Cross 1822.

E 11 : POLLY—Pansy 1063. Patchwork 2428. Penelope of Elmham 2437. Priscilla of Elmham 2472.

L 3 : ELMER—Elm-branch 1482. Emma of Elmham 2164.

N 6 : TIT—Rinaldo 736. Cherry Bloom 1381. Cherry of Elmham 2082. Crop-ears 2115.

O 14 : CHERRY—Charlotte 2068.

P 3 : ROSE—Thursford Rose 600.

T 7 : STRANGER—Songster 1859. Sonsie of Elmham 2541.

1 Norf.: POND—Pond Lily 2451. Pond Lily 2nd 2452. Pond Lily 3rd 2453. Wiffen 2605.

PROBATIONERS :—Baly 2nd 1995. Blyth 2033. Clarke 2nd 2095. Madame Nicholson 2345.

SUMMARY :—Of the Elmham Group 6 Bulls 14 Cows. Biddell Group 1 Cow. Eaton Group 5 Cows. Dereham Group 2 Cows. Necton Group 1 Bull 3 Cows. Oakley Group 1 Cow. Powell Group 1 Cow. Thursford Group 2 Cows. Norfolk Group 4 Cows. Probationers 4 Cows. Total 44.

HERDS AND OWNERS.

REV. J. C. GIRLING,

Hautbois Rectory, Norfolk.   Post Town—Norwich.

A 22 : ALICE—Nosegay 2417.
A 23 : DIDO—Patty 2431.
E 3 : COUNTESS—Eaton 666.

SUMMARY :—Of the Elmham Group 2 Cows. Eaton Group 1 Bull. Total 3.

MR. G. GOODERHAM,

Monewden, Suffolk.   Post Town—Wickham Market.

V 1 : COWSLIP—Dallinghoo 650. Wild Ruler 779. Wild Rhona 2608. Wild Rose 1271. Wild Rose of Kilburn 1939. Wild Rosy 2609.
V 2 : RED STOCKINGS—Troston 6th 796. Troston 7th 797. Floss 2nd 1523. Wild Cherry 2607.
V 3 : FILLPAIL—Fancy 1598. Fairy 2179.
1 Suff. : BAKER—Baker 1992. Miss Baker 2377.

SUMMARY :—Of the East Suffolk Group 4 Bulls 8 Cows. Suffolk Group 2 Cows. Total 14.

BAZETT M. HAGGARD, Esq.,

Kirby Cane Hall, Norfolk.   Post Town—Bungay.

B 2 : CHERRY LUX—Sandboy 747.
H 1 : DAVY—Davyson 10th 479.
K 19 : ROSE—Young Rose 1593.
N 4 : ROSE—Fleck 2192. Marguerite 1657. Marguerite 2nd 1658. Marguerite 3rd 2355. Rose 1142. Winter-blossom 1940.
N 6 : TIT—Judy 1602. Judy 2nd 2278. Tit 3rd 1898.

HERDS AND OWNERS.

MR. HAGGARD'S HERD—*Continued.*

R 9 : BRUNDISH—Pretty 2463.

   SUMMARY—Of the Biddell Group 1 Bull. Hammond Group 1 Bull. Kimberley Group 1 Cow. Necton Group 9 Cows. Starston Group 1 Cow. Total 13.

HER GRACE THE DUCHESS OF HAMILTON AND BRANDON,

Easton Park, Suffolk. Post Town—Wickham Market.

Steward—Mr. D. SMITH.

A 4 : RINGLET—JUMBO 683. Kattie 975. Katie's Sister 1604. Little Katie 1630.

B 8 : HANDSOME—Lady Handsome 2292.

B 12 : BEE—Honeywood 2256. Little Bee 1003.

E 2 : CHERRY—DONCASTER 661. MADCAP 697. Radish 1761. Theodosia 1227.

I 2 : RUBY—BUSTLE 635. Jessie 961.

N 2 : MINNIE—Lily 2314. Milly 2nd 1670.

O 2 : QUEEN—Ruby 1164.

O 7 : DAISY—Daisy 2119. Daisy Girl 1439.

V 14 : HONESTY—Glemham Rose 921. Rosebud 1156. Rosy 1812.

W 14 : CLARA—Bracelet 2037. Easton Gem 868. Emerald 204. Esmeralda 873. The Gem 2208.

X 3 : COSSET—Camelia 742. Camelia Bud 743. Blossom 2027.

X 5 : BLUE BELL—Blue Bell 713. Harebell 2241. Miss Bell 2378.

   SUMMARY :—Of the Elmham Group 1 Bull 3 Cows. Biddell Group 3 Cows. Eaton Group 2 Bulls 2 Cows. Hudson Group 1 Bull 1 Cow. Necton Group 2 Cows. Oakley Group 3 Cows. East Suffolk Group 3 Cows. Wolton Group 5 Cows. Trimley Group 6 Cows. Total 32.

MR. JOHN HAMMOND,

Bale, Norfolk.   Post Town—East Dereham.

H 1 : DAVY—Davyson 7th 476. Davyson 11th 480. Davyson 15th 652. Davyson 16th 653. Davy 5th 167. Davy 7th 169. Davy 15th 844. Davy 18th 847. Davy 21st 1445. Davy 22nd 1446. Davy 23rd 1447. Davy 25th 1449. Davy 27th 1451. Davy 28th 1452. Davy 29th 1453. Davy 32nd 1456. *Davy 33rd 1457. Davy 34th 1458. Davy 35th 1459. Davy 36th 2129. Davy 37th 2130. Davy 38th 2131. Davy 39th 2132. Davy 41st 2133. Davy 42nd 2134. Davy 43rd 2135. Davy 44th 2136. Davy 45th 2137. Davy 46th 2138. Davy 47th 2139. Davy 48th 2140. Davy 49th 2141. Davy 50th 2142. Davy 51st 2143.

H 2 : BUTLER—Beauty 26. Beauty 3rd 1309.

Summary:—Of the Hammond Group 4 Bulls 32 Cows. Total 36.

* See Corrected Entry.

T. HARRISON, Esq.,

Copford Hall, Essex.   Post Town—Colchester.

O 3 : COWSLIP—Copford Rose 2101.

O 8 : MARY GREY—Juncate 2282.

R 2 : LOVELY—Copford Prince 644.

Summary:—Of the Oakley Group 2 Cows. Starston Group 1 Bull. Total 3.

SIR J. W. C. HARTOPP, Bart.,

Kingswood Warren, Surrey.   Post Town—Epsom.

Steward—Mr. Andrew Gordon, The Warren Farm, Kingswood.

A 2 : CHERRY—Yellow Girl 1948.

A 5 : RAMSLEY—Greenfield 2221. Jessie 2269. Mrs. Bradfield 1686.

SIR J. W. C. HARTOPP'S HERD—*Continued.*

A 13 : SPOT—Annie 1984. Daisy 2218. Jennie 2265. Red Stocking 1778.

E 11 : POLLY—Annie 1985. Gladiolus 1537. Glunberry 2219.

G 13 : GOLDEN DROP—Castleacre 1364. Clarea 2092. Congress 2099. Constance 2100.

K 17 : CHERRY—Hardwick 501. Annie 1986. Bright Cherry 724. Powly 2460.

O 1 : DUCHESS OF SUFFOLK—Pink 2446. Princess Royal 1755.

O 6 : VANITY—Canteen 1361. Fannie 2185. Nanzie 2403. Susie 2564.

P 3 : ROSE—Bounty 460.

P 4 : NINA—Beatrice 690. Gloss 2216. Milkmaid 2367.

S 2 : HAPTON CHERRY—Beauty 2006. Fillpail 2188. Flora 2194. Red-berry 1765. Rose 2499.

SUMMARY :—Of the Elmham Group 8 Cows. Eaton Group 3 Cows. Castleacre Group 4 Cows. Kimberley Group 1 Bull 3 Cows. Oakley Group 6 Cows. Powell Group 1 Bull 2 Cows. Stoke Group 6 Cows. Total 34.

MR. R. L. HARVEY,

Wyken Hall, Bardwell, Suffolk.   Post Town—Ixworth.

U 30 : RINGLET—Valour 774.

MR. W. HARVEY,

Timworth, Suffolk.   Post Town—Bury St. Edmund's.

V 5 : CHERRY—Hamlet 500.

U 23 : ANNIE—Annie 681. Olga 2424. Queen 2478. Juliet 2281.

U 25 : DAHLIA—Dahlia 818.

MR. HARVEY'S HERD—*Continued.*

U 26 : DAIRYMAID—Dairymaid 2nd 1429. Lady Macbeth 2297.

U 27 : DAISY—Nancy Lee 1961.

U 29 : RED ROSE—Princess Margaret 2471. Red Rose 1126.

U 31 : ROSAMOND—Ophelia 2426. Rosamond 1139.

U 32 : ROSEBUD—Portia 2459. Rosebud 2nd 1806.

U 34 : RUTH—Ruth 1170.

SUMMARY :—Of the East Suffolk Group 1 Bull 2 Cows. West Suffolk Group 13 Cows. Total 16.

THE RIGHT HON. LORD HASTINGS,

Melton Constable, Norfolk. Post Town—East Dereham.

Steward—Mr. E. W. BECK.

E 5 : ROSE—ROSCOE 559.

H 1 : DAVY—BARON ROSCOE 621. MELTON DAVID 699. RUPERT 746. YOUNG DUKE 781. Baroness Davy 1997. Davy 16th 845. Davy 17th 846. Davy 19th 848. Davy Duchess 1460. Davy Duchess 2nd 2144. Davy Duchess 3rd 2145. Ladyday 2289. Melton Davy 1663. Melton Davy 2nd 2362. Melton Davy 3rd 2363. Thornham Davy 1890. Thornham Davy 2nd 1891. Thornham Davy 2571.

K 18 : CHARMER—Charmer 4th 2070.

K 19 : ROSE—FRITZ 671. Fillpail 1512.

L 2 : HAWKEYES—Hawkeyes 5th 261.

N 6 : TIT—Dainty 819. Nectarine 2404. Nectarine 2nd 2405.

P 6 : NANCY—Lily 2316.

P 7 : VIOLET—Melton Rose 2364. Melton Rose 2nd 2365. Rosebud 1804.

SUMMARY :—Of the Eaton Group 1 Bull. Hammond Group 4 Bulls 14 Cows. Kimberley Group 1 Bull 2 Cows. Dereham Group 1 Cow. Necton Group 3 Cows. Powell Group 4 Cows. Total 30.

MR. HENRY HAYLOCK,

Saham Grove, Norfolk.   Post Town—Shipdham.

A 1 : PRIMROSE—Blossom 2023. Helen 2246.

E 2 : CHERRY—Rebecca 1763.

N 6 : TIT—Edith 2158. Rose 2496.

V 14 : HONESTY—The Pope 727.

   Summary :—Of the Elmham Group 2 Cows. Eaton Group 1 Cow.
   Necton Group 2 Cows. East Suffolk Group 1 Bull. Total 6.

The Right Hon. LORD HENNIKER,

Thornham Hall, Suffolk.   Post Town—Eye.

Steward—Mr. W. H. Preston.

A 19 : LADY CONSTABLE—Lady Constable 300. Twin Sister
        A 2580. Twin Sister B 2581.

K 17 : CHERRY—Fairy Queen 1503. Princess C 2469. Queen D
        2476. Queen G 2477. Thornham Princess 1230.

O 11 : POLLY—Cyprus 2nd 648. Polly 1085.

O 12 : BEAUTY—Lord Hartismere 695. Sprightly 2nd 1201.

O 13 : STRAWBERRY—Thornham Hero 763. Daisy 828. Pearl
        1727. Ruby 1165. Ruby A 2518. Strawberry 2nd 1212.
        Strawberry K 2553. Pearl I 2435.

O 14 : CHERRY—Rose Crown 1807. Rose E 2497. Rose F 2498.
        Rose-leaf 1159. Rose-leaf 2nd 1816. Rose-leaf B 2506.
        Rose-leaf H 2507.

   Summary :—Of the Elmham Group 3 Cows. Kimberley Group 5
   Cows. Thornham Group 3 Bulls 16 Cows. Total 27.

HERDS AND OWNERS.

GEORGE HOLMES, Esq.,
Brooke, Norfolk.   Post Town—Norwich.

R 8 : BEAUTY—Quality 2475.

R 9 : BRUNDISH—Swipes 761. Beauty 2005.

S 3 : DOWSON—Brundish Prince 462.

U 45 : CONSTANCE—Constance 3rd 1399. Constance 6th 1401.

W 10 : TOPKNOT—Hunston Topknot 1579.

   Summary:—Of the Starston Group 1 Bull 2 Cows. Stoke Group 1 Bull. West Suffolk Group 2 Cows. Wolton Group 1 Cow. Total 7.

MR. JOHN HOWELL,
Great Walsingham, Norfolk.   Post Town—Walsingham.

T 15 : DAPHNE—Buttercup 4th 1347. Buttercup 5th 1348. Buttercup 6th 1342. Cowslip 2nd 1416. Cowslip 3rd 2111.

T 17 : ABBESS—Friday 2nd 1526. Friday 3rd 1527. Friday 4th 2201. Friday 5th 2202. Friday 6th 2203. Handsome 1553. Handsome 2nd 2229.

T 18 : NUN—Doctor Davyson 658. Nancy 1039. Nancy 2nd 1693. Nancy 3rd 1694. Nancy 4th 1695. Nancy 7th 2396. Nancy 8th 2397.

   Summary:—Of the Walsingham Group 1 Bull 18 Cows. Total 19.

MR. JOHN HOWLING,
Elmham, Norfolk.   Post Town—East Dereham.

A 12 : HANDSOME—Boulter 54. Bounce 2035. Nancy 2392. Nan of Elmham 2391.

A 13 : SPOT—Sprightly 2546.

   Summary:—Of the Elmham Group 5 Cows.

MR. WILLIAM HUDSON,

Quarles, Norfolk.   Post Town—Fakenham.

I 16 : CORAL—Countess 1406.   Crocus 2114.   Handsome 2228.

I 17 : ISABELLA—Isabella 1589.   Isabella 3rd 1591.   Isabel 2261.

I 18 : LUCY—Lady Bird 2288.   Laurel 2306.   Little Lady 2319. Lucy 1st 1642.   Lucy 2nd 1643.   Lucy 3rd 1644.   Lucy 4th 1645.   Lucy 5th 1646.

I 19 : MARGARET—Maggie 2347.   Margaret 1654.   Margaret 2nd 1655.

I 20 : RUDDY—Rose 1792.

I 21 : SEPSY—Jessie 2270.   Lady Mabel 2296.

I 22 : SPRIGHTLY—Sprightly 1864.   Sprightly 2nd 1865.

I 23 : SUSAN—Susan 1st 1878.   Susan 2nd 1879.   Susie 2563.

K 24 : FLORA—Flo 2193.   Flora 1520.   Flora 2nd 1521.

S   2 : HAPTON CHERRY—Haman 499.

SUMMARY:—Of the Hudson Group 25 Cows. Kimberley Group 3 Cows. Stoke Group 1 Bull. Total 29.

MESSRS. D. AND G. JONES,

Galesburg, Illinois, U.S.A.

AU 5 : PRIMULA—Commander 642.

MR. F. D. KENT,

Corringham, Essex.   Post Town—Romford.

B   7 : LILY—Water Lily 644.

B   9 : ROSE—June Rose 968.   Juno 2283.   Rose 1791.

B 17 : FAIRY—Wiseacre 780.   Wise Princess 2611.   Wisdom 656.

V   5 : CHERRY—Mint 1676.

HERDS AND OWNERS.

MR. KENT'S HERD—*Continued.*

V 10 : GRIMACE—Gain 1533.

W 14 : CLARA—Enchantress 2166. Witch 657.

SUMMARY :—Of the Biddell Group 1 Bull 6 Cows. East Suffolk Group 2 Cows. Wolton Group 2 Cows. Total 11.

SIR E. C. KERRISON, BART.,

Oakley Park, Norfolk.   Post Town—Scole.

Steward—Mr. C. H. SCRIVEN, Oakley House.

E 5 : ROSE—* ROMANO 740.

L 12 : KATE—Kate 971. Kate 2nd 2285.

O 1 : DUCHESS OF SUFFOLK—Cherry 2nd 2078. Duchess of Suffolk 2nd 2155. Minnie 2nd 1675. Oakley 4th 2423. Victoria 2598.

O 5 : NOSEGAY—Nosegay 3rd 1055. Nosegay 4th 2418. Nosegay 5th 2419.

O 6 : VANITY—Eyebright 2nd 879. Eyebright 3rd 2173. Mirth 2nd 1678. Mirth 3rd 2374. Mirth 4th 2375.

O 8 : JEWEL—Jewel 2nd 2272. Jewess 2273.

P 1 : HANDSOME—Countess 2105.

R 2 : LOVELY—PASTOR 715.

SUMMARY :—Of the Eaton Group 1 Bull. Dereham Group 2 Cows. Oakley Group 15 Cows. Powell Group 1 Cow. Starston Group 1 Bull. Total 20.

* See Corrected Entry.

THE RIGHT HON. THE EARL OF KIMBERLEY,

Kimberley House, Norfolk.   Post Town—Wymondham.

Bailiff—Mr. KING.

K 1 : CROWNTHORPE—Crownthorpe 4th 812. Crownthorpe 5th 1420. Crownthorpe 6th 1421.

HERDS AND OWNERS.

The Right Hon. THE EARL OF KIMBERLEY'S HERD
Continued.

K 3 : DAISY—Daisy 151. May 1960. Pansy 1721. Violet 1251.

K 4 : BUTTERCUP—Anemone 676. Crocus 1417. Hyacinth 1583. Lily 1626.

K 5 : YOUNG GAP—Tulip 616. Queen 1118. Snowdrop 1193. Sunflower 1877.

K 6 : PRIMROSE—Cowslip 1414. Princess 1106.

K 7 : NANCY—Beauty 694. Polly 1083. Topsy 1237.

P 4 : NINA—Adonis 615.

SUMMARY:—Of the Kimberley Group 20 Cows. Powell Group 1 Bull. Total 21.

MR. J. M. KNAPP,
Bellevue, Michigan, U.S.A.

AA 29 : BELLE—Prince Albert 729.

MR. P. LEEDER,
Brooke, Norfolk.   Post Town—Norwich,

E 11 : POLLY—Grasshopper 674.

REV. A. G. LEGGE,
Elmham, Norfolk.   Post Town—East Dereham.

A 11 : NANCY—Fan of Elmham 2182. Fortuna 1524. Frances 904. Frolic 906.

A 27 : CURSON—Curson 813. Curson Cherry 2116.

SUMMARY:—Of the Elmham Group 6 Cows.

HERDS AND OWNERS.

MR. J. T. LEWIS,

Chatham, Sangamon Co., Illinois, U.S.A.

AA 12 : HANDSOME—Belle of Chatham 2011.

AU 6 : PHŒNIX—Superb 759.

Summary :—Of the American Group 1 Bull 1 Cow.

R. E. LOFFT, Esq.,

Troston Hall, Suffolk.   Post Town—Bury St. Edmund's.

Steward—Mr. Jeavons.

A 1 : PRIMROSE—Elmham Nelly 2nd 1492. Elmham Nelly 3rd 2160. Elmham Rosebud 2nd 872. Elmham Rosebud 3rd 2161. Elmham Rosebud 4th 2162.

A 3 : BRIGHT—Dale 649. Elmham 2nd 1484.

B 10 : BURY—Silver Locks 551. Silverlocks 2nd 2538.

I 9 : BRIDESMAID—Bridegroom 2nd 630. Bridesmaid 2nd 721. Bridesmaid 3rd 722. Bridesmaid 4th 1334. Bridesmaid 5th 1335. Bridesmaid 6th 1336. Bridesmaid 7th 1337. Bridesmaid 8th 2039.

I 13 : ROSEBUD—No Doubt 707. Smart 757. Rosebud 492. Rosebud 2nd 1153. Rosebud 3rd 1798. Rosebud 4th 1799. Rosebud 5th 1800. Rosebud 6th 1801. Rosebud 7th 1802.

M 2 : RED ROSE—Helena 944. Helena 2nd 1565. Helena 3rd 2247.

N 2 : MINNIE—Lily 3rd 1000. Minnie 3rd 343. Minnie 5th 1673. Minnie 6th 1674. Minnie 7th 2368. Minnie 8th 2369. Minnie 9th 2370.

U 2 : FLOSS—Dolly 179. Dolly 2nd 1465. Dolly 3rd 1466.

U 3 : HANDSOME—Stout 581. Handsome 3rd 250. Handsome 4th 934. Handsome 5th 935. Handsome 6th 936. Handsome 8th 1554. Handsome 9th 1555. Handsome 10th 1556. Handsome 11th 2230. Handsome 12th 2231. Handsome 13th 2232. Handsome 14th 2233. Handsome 15th 2234.

MR. LOFFT'S HERD—*Continued.*

U 5 : PRIMULA—Cauliflower 4th 755.

U 6 : PHŒNIX—Daisy 156. Daisy 2nd 2121. Phœnix 2nd 2442.

U 9 : WAXWORK—Slasher 577. Waxwork 2nd 648. Waxwork 4th 1930.

U 43 : POPPET—Powerful 728. Poppet 1086. Poppet 2nd 1087. Poppet 3rd 1742. Poppet 4th 1743. Poppet 5th 2454.

V 11 : GLOSS—Small 756. Gloss 2nd 665. Gloss 3rd 1542. Gloss 4th 1543. Gloss 5th 2217. Gloss 6th 2218.

W 3 : NELLY—A Live Bull 617. Newbourn Pride 3rd 1050. Newbourn Pride 4th 1051. Newbourn Pride 6th 1707. Newbourn Pride 7th 1708. Newbourn Pride 8th 1709. Newbourn Pride 9th 1710. Newbourn Pride 10th 1711. Newbourn Pride 11th 2409. Wide-awake 655. Wide-awake 3rd 1936. Wide-awake 4th 1937.

W 10 : TOPKNOT—Topknot 3rd 1908.

Summary :—Of the Elmham Group 1 Bull 6 Cows. Biddell Group 2 Cows. Hudson Group 3 Bulls 14 Cows. Marham Group 3 Cows. Necton Group 7 Cows. West Suffolk Group 3 Bulls 26 Cows. East Suffolk Group 1 Bull 5 Cows. Wolton Group 1 Bull 12 Cows. Total 84.

REV. H. EVANS-LOMBE,

Bylaugh Hall, Norfolk.   Post Town—East Dereham.

Steward—Mr. E. A. Limmer.

A 5 : RAMSLEY—Lord Bylaugh 3rd 692. Blossom 2024.

A 20 : FUCHSIA—Lily 1624.

F 4 : SNELLING—Lizzie 1634.

F 6 : CLARA—Lilac 1621.

I 12 : COWSLIP—Lively 1632. Lively 2nd 2312. Louie 2327. Lovely 1636.

K 19 : ROSE—Lord Bylaugh 2nd 691. Joyful 1598.

HERDS AND OWNERS.

REV. H. EVANS-LOMBE'S HERD—*Continued.*

L 11 : LETTON—Lassie 2301. Leaf 2307. Leech 2308. Lida 2310. Likely 2312. Linda 2317. Lottie 2325. Louise 2328. Lucky 2334.

N 6 : TIT—Bylaugh Tit 2052. Bylaugh Tit 2nd 2053. Tit 2nd 1897.

W 3 : NELLY—Lady 2287.

W 15 : DAISY—Laura 1615.

PROBATIONER—Little 2318.

SUMMARY :—Of the Elmham Group 1 Bull 2 Cows. Easton Group 2 Cows. Hudson Group 4 Cows. Kimberley Group 1 Bull 1 Cow. Dereham Group 9 Cows. Necton Group 3 Cows. Wolton Group 2 Cows. Probationer 1 Cow. Total 26.

LIEUT.-COL. W. BEESTON LONG,

Hurts Hall, Suffolk.   Post Town—Saxmundham.

V 15 : ALICE—Agnes 1979. Alicia 673.

V 16 : RUBY—Rose 1794. Ruby 467. Russet 1834. Ruth 2523.

V 17 : UNA—Honey 950. Honey-dew 2255. Honeymoon 1573. Honeysuckle 1574. Vanity 1922.
BUTLEY 465.

PROBATIONER :—Rachel 1760.

SUMMARY :—Of the East Suffolk Group 11 Cows. Wolton Group 1 Bull. Probationer 1 Cow. Total 13.

J. J. L. LUBBOCK, ESQ.,

Catfield Hall, Norfolk.   Post Town—Norwich.

A 20 : FUCHSIA—Lucy 1010. Lucy Glitters 2339. Lucy Jane 2340.

M 5 : SYBIL—Laurel Wreath 1617. Lauristina 995.

SUMMARY :—Of the Elmham Group 3 Cows. Marham Group 2 Cows. Total 5.

MR. FAIRMAN J. MANN,

Shropham, Norfolk.   Post Town—Thetford.

T   4 : TIT—Brummell 632.

2 Norf.: MANN—Magdalen 2346. Maid Marian 2349. Maid Marian 2nd 2350. Maid Marian 3rd 2351. Margery 2354. Maria 2358. Mattie 2360. Missie 2376. Miss Maria 2380. Miss Mattie 2381.

Summary:—Of the Thursford Group 1 Bull. Norfolk Group 10 Cows. Total 11.

MR. JOHN MARGARSON,

Wendling, Norfolk.   Post Town—East Dereham.

H   1 : DAVY—Davyson 14th 651.

L   3 : ELMER—Uncas 772. Ultra 2584. Una 1243. Una 2nd 1920. Uniform 2585. Unison 2586. Upland 2587. Upton 1245. Upton 2nd 1921. Upshot 2588. Urbane 2589.

L   9 : CHERRY—Cassie 2058. Chalice 2064. Charmer 2071. Cheerful 763. Cheerly 1370. Cherry 2nd 1375. Cherry 3rd 1376. Cloister 2096. Comfit 2098.

U 43 : POPPET—John 510.

Summary:—Of the Hammond Group 1 Bull. Dereham Group 1 Bull 19 Cows. West Suffolk Group 1 Bull. Total 22.

J. L. MARRIOTT, Esq.,

Narborough, Norfolk.   Post Town—Swaffham.

A   4 : RINGLET—Cassandra 2057. Cressida 2113. Kitten 1608.

Summary:—Of the Elmham Group 3 Cows.

R. HARVEY MASON, Esq.,

Necton Hall, Norfolk.   Post Town—Swaffham.

Steward—Mr. W. Fowler.

A 33 : ELM-LEAF—Cellarette 2063. Elvira 2163. Nancy 1688. Nina 2412.

I 9 : BRIDESMAID—Orlando 712.

I 13 : ROSEBUD—Rosebud 7th 1802.

L 9 : CHERRY—Marchioness 2352. Margery 2353.

N 1 : DARLING—Petitioner 718. Petrarch 719. Pet 1072. Phœbe 1733.

N 4 : ROSE—Prince Imperial 731. Empress 1496. Eugenie 1499. Eulalie 1499. Frances 2198. Phillippa 2439. Raspberry 2482. Rosalind 1787. Rose 2nd 1143. Rosebud 1803. Rosina 2510. Rosinette 2511. Starling 2550. Strawberry 2nd 2552. Tulip 2nd 1241.

N 5 : TULIP—Semiramis 2527. Sheba 1841. Sultana 1876. Zenobia 2612.

N 6 : TIT—Tancred 762. Victor 773. Dainty 1427. Darling 1443. Heroine 2251. Viola 2600. Violet 4th 1252. Victoria 1926.

N 7 : SKELTON—Lupin 1648. Lotus 2326.

R 2 : LOVELY—Starston Duke 570.

Summary:—Of the Elmham Group 4 Cows. Hudson Group 1 Bull 1 Cow. Dereham Group 2 Cows. Necton Group 5 Bulls 28 Cows. Starston Group 1 Bull. Total 42.

COL. JNO. B. MEAD and ROBERT J. KIMBALL, Esq.,

Suffolk Farm, Randolph, Orange Co., Vermont, U.S.A.

A 1 : PRIMROSE—Bloom 1325.

A 4 : RINGLET—Duchess of Hamilton 2154.

COL. MEAD'S AND MR. R. J. KIMBALL'S HERD
Continued.

AA 11 : NANCY—Pomp 541.

E 11 : POLLY—Polly 2nd 1738.

H 1 : DAVY—Davy Princess 2146.

H 2 : BUTLER—Beauty 4th 1310.

I 9 : BRIDESMAID—Vermont Bridesmaid 2593.

P 3 : ROSE—Romeo 741. Ruby Rose 1830.

U 2 : FLOSS—Flossie 2196.

W 2 : BEAUTY—Winsome 1943.

W 10 : TOPKNOT—Vermont Topknot 2594.

SUMMARY:—Of the Elmham Group 2 Cows. American Group 1 Bull. Eaton Group 1 Cow. Hammond Group 2 Cows. Hudson Group 1 Cow. Powell Group 1 Bull 1 Cow. West Suffolk Group 1 Cow. Wolton Group 2 Cows. Total 12.

MR. FREDERICK MORRIS,

Hall Farm, Geldeston, Norfolk.   Post Town—Beccles.

E 11 : POLLY—Bridesmaid 2038.

SUMMARY:—Of the Eaton Group 1 Cow.

MR. J. L. MUSTARD,

Lebanon, Missouri, U.S.A.

AN 7 : SKELTON—Nora 2416.

AO 12 : BEAUTY—Judge 680.

SUMMARY:—Of the American Group 1 Bull 1 Cow.

HERDS AND OWNERS.

MR. GEORGE JNO. PAINE,

Risby, Suffolk.   Post Town—Bury St. Edmund's.

U 48 : VISCOUNTESS—Viscountess 2nd 1928.   Viscountess 3rd 2602.

W 10 : TOPKNOT—Risby Topknot 2491.

W 21 : CHLOE—Chloe 6th 2088.

SUMMARY:—Of the West Suffolk Group 2 Cows. Wolton Group 2 Cows. Total 4.

MR. T. LEONARD PALMER,

Banham, Norfolk.   Post Town—Attleborough.

A 3 : BRIGHT—Elmham 3rd 1485.   Pimpernel 2444.

K 19 : ROSE—Alonso 447.

K 23 : KATE—Charity 2067.   Chaste 1369.   Cherry 1374.

N 2 : MINNIE—Prospero 732.   Lily 4th 1627.   Thrift 2573.

U 14 : SWEET PEA—Timon 765.   Sweet Pea 2567.   Tansy 2570.

U 27 : DAISY—Woodbine 1947.

U 34 : RUTH—Beauty 1313.   Lupine 2341.

SUMMARY:—Of the Elmham Group 2 Cows. Kimberley Group 1 Bull 3 Cows. Necton Group 1 Bull 2 Cows. West Suffolk Group 1 Bull 5 Cows. Total 15.

MRS. PERKINS,

Saham Hall, Norfolk.   Post Town—Watton.

Agent—Mr. J. ALBRECHT, Rokeles Hall, Watton.

A 33 : ELM-LEAF—Ash-leaf 1989.   Ash-leaf 2nd 1990.   Elm-leaf 2nd 1489.   Elm-leaf 3rd 1490.   Handsome 1551.   Ivy 2262.

HERDS AND OWNERS.

MRS PERKINS' HERD—*Continued.*

K 19 : ROSE—OTHELLO 532.

N 1 : DARLING—ARABI 619. Blossom 1328. Blossom 2nd 2026.

N 4 : ROSE—THE COLONEL 642. Beauty 2004. Daisy 3rd 1434. Rosette 1819. Rosette 2nd 1820. Rosette 3rd 2502. Rosette 4th 2503. Rosette 5th 2504. Strawberry 1872.

N 5 : TULIP—Red Tulip 1771. Red Tulip 2nd 2489.

N 10 : VIOLET—Violet 1925.

SUMMARY :—Of the Elmham Group 6 Cows. Kimberley Group 1 Bull. Necton Group 2 Bulls 13 Cows. Total 22.

MR. N. POWELL,

Glandford, Norfolk.   Post Town—Cley-next-the-Sea.

P 5 : RASPBERRY—HECTOR 675. Hetty 1570.

P 6 : NANCY—PLOUGHBOY 725. Laura 2305. Lilian 1623. Lily of Glandford 2315. Lively 1633. Lucy 2nd 2337. Priscilla 2nd 1965.

P 7 : VIOLET—GLANDFORD PRINCE 673. PRIME MINISTER 545. ROLAND 739. ROYAL DUKE 743. RUPERT 745. Polly 2448. Primrose 3rd 1749. Rose-leaf 1817. Ruby 1825.

P 9 : CHERRY—Careless 2056. Cherry 2nd 1377. Comassie 1957. Countess 2106. Cowslip 2110. Daisy 2nd 1437. Mabel 2344. Maggie 2348. Marian 1659. Mermaid 1665. Mildred 1667. Milkmaid 1668. Minnie 2371. Molly 1682. Moss Rose 2nd 1684. Nancy 3rd 2393.

SUMMARY :—Of the Powell Group 7 Bulls 27 Cows. Total 34.

C. S. READ, ESQ.,

Honingham Thorpe, Norfolk.   Post Town—Norwich.

A 31 : STAR—Sybil 1st 1884. Sybil 2nd 1885. Sybil 3rd 1886. Sybil 4th 1887. Sybil 5th 1888. Sybil 6th 2568. Sybil 7th 2569. Sylph 1223.

MR. READ'S HERD—*Continued.*

F 9 : BOO—Bones 715. Boo 1st 1329. Boo 2nd 1330. Boo 3rd 1331.

F 10 : COO—Coo 802. Coo 1st 1402. Coo 1403. Coo 3rd 1404.

SUMMARY :—Of the Elmham Group 8 Cows. Easton Group 8 Cows. Total 16.

MR. JAS. RIVETT,

Mileham, Norfolk.   Post Town—East Dereham.

A 1 : PRIMROSE—NIPPER 705.

1 Norf.: POND—Charlotte 2069. Cherry 2080. Dairymaid 2117. Damson 2125. Dancer 2126. Dapple 2127. Jemima 2264. Jenny 2266. Lucy 2338. Nancy 2399. Nelly 2406. Nest 2408. Nutmeg 2422. Polly 2450. Poppy 2457. Prude 2473.

SUMMARY :—Of the Elmham Group 1 Bull. Norfolk Group 16 Cows. Total 17.

JOHN F. ROGERS, ESQ.,

Swanington, Norfolk.   Post Town—Norwich.

A 1 : PRIMROSE—Agatha 1302. Amethyst 1305.

E 3 : COUNTESS—Esther 2nd 2169.

E 12 : SUSAN—EMPEROR 489. Eglantine 1476. Empress 1495. Endive 1497. Eugenie 2170. Eurotas 2172. Eyebright 1502. Susanna 2nd 2562.

F 8 : FANNY—FESTUS 668. Fairy 2nd 2178. Fanny 899. Flame 2190. Foxglove 1525.

SUMMARY :—Of the Elmham Group 2 Cows. Eaton Group 1 Bull 8 Cows. Easton Group 1 Bull 4 Cows. Total 16.

GENERAL L. F. ROSS,

Iowa City, Iowa, U.S.A.

A 12 : HANDSOME—Ocean Maid 401.

AA 29 : BELLE—Dexter 654.

V 2 : RED STOCKINGS—Red Beauty 2483.

V 14 : HONESTY—Jilt 2274.

> Summary :—Of the Elmham Group 1 Cow. American Group 1 Bull. East Suffolk Group 2 Cows. Total 4.

GEORGE L. SANDERSON, Esq.,

Nippenose Park Farm, Williamsport, Pennsylvania, U.S.A.

V 1 : COWSLIP—Wild Rolla 604.

AA 12 : HANDSOME—Nippenose Belle 2413.

> Summary :—Of the East Suffolk Group 1 Bull. American Group 1 Cow.

MR. A. J. SMITH,

Rendlesham, Suffolk.   Post Town—Woodbridge.

B 11 : SUFFOLK—Belle 2010. Eyke Belle 2174. Eyke Duchess 2175. Eyke Lassie 2177. Suffolk Duchess 2557. Ufford Belle 2582. Ufford Duchess 2583.

B 13 : BLOSSOM—Black Blossom 2020. Bushy Blossom 2047. Peach Blossom 2432. Peach Bud 2433. Peach Leaf 2434. Red Blossom 2485.

B 18 : FANCY—Fine Fruit 2189. Fresh Fruit 2200.

B 19 : GIPSY—Eyke Gipsy 2176.

B 20 : PICKET—Brown Loo 2043. Knock-in 2286. Loo 2324.

B 21 : ROSEBUD—Spring Leaf 2549. Summer Leaf 2558.

K 17 : CHERRY—Princess Cherry 1753.

MR. SMITH'S HERD—*Continued.*

O 13 : STRAWBERRY—Ruby Crown 1829.
O 14 : CHERRY—Cherry Rose 2087.
V 5 : CHERRY—Bold Heart 626.
V 14 : HONESTY—Pickwick 720.

SUMMARY :—Of the Biddell Group 21 Cows. Kimberley Group 1 Cow. Oakley Group 2 Cows. East Suffolk Group 2 Bulls. Total 26.

MR. J. McLAIN SMITH,

Dayton, Ohio, U.S.A.

AA 29 : BELLE—Duke of Dayton 663.

MR. D. L. STEVENS,

Elkdale, Susquehanna Co., Pennsylvania, U.S.A.

AA 1 : PRIMROSE—Diana 2147.
AA 13 : SPOT—Spinster 2nd 2544.
AN 7 : SKELTON—Rinaldo 737.

SUMMARY :—Of the American Group 1 Bull 2 Cows.

The Right Hon. W. H. SMITH, M.P.,

Great Thurlow, Suffolk.   Post Town—Newmarket.

Steward—Mr. Andrew Boa.

B 9 : ROSE—Brownlow 631.
B 11 : SUFFOLK—Aconite 1976. Belliza 1319.
I 9 : BRIDESMAID—Daisy 2119.
I 13 : ROSEBUD—Buttercup 2050. Goldie 2220.
N 2 : MINNIE—Winnie 2610.

HERDS AND OWNERS.

THE RIGHT HON. W. H. SMITH'S HERD—*Continued.*

W 3 : NELLY—Little Nell 2320. Snowdrop 2540.

SUMMARY—Of the Biddell Group 1 Bull 2 Cows. Hudson Group 3 Cows. Necton Group 1 Cow. Wolton Group 2 Cows. Total 9.

MR. J. M. SPINKS,

Harpley, Norfolk.   Post Town—Rougham.

P 2 : STRAWBERRY—DON PEDRO 660.

V 9 : GLAD—Glad 920. Gleaner 2213. Glee 2nd 663. Glide 2214.

V 10 : GRIMACE—Gad 2205. Gadfly 1532. Gainful 2206. Grim 2223. Grimace 4th 927. Grimace 5th 928. Grimace 6th 1959.

V 11 : GLOSS—Penal 2436. Penance 1728. Pride 2466. Princess 1109.

V 13 : ROWLEY—Garnet 2207. Gland 1589. Glemham 923. Glimmer 2215. Grace 925. Rowley 2515. Ruby 1827. Rustic 2521.

SUMMARY :—Of the Powell Group 1 Bull. East Suffolk Group 23 Cows. Total 24.

MR. HY. SPURLING,

Shotley, Suffolk.   Post Town—Ipswich.

U 16 : VIOLET—Ampton 1982.

V 11 : GLOSS—SPARK 579.

SUMMARY :—Of the West Suffolk Group 1 Cow. East Suffolk Group 1 Bull.

MR. G. P. SQUIRES,

Marathon, New York, U.S.A.

A 11 : NANCY—FRANK OF ELMHAM 670.

HERDS AND OWNERS.

MR. SQUIRES' HERD—*Continued.*

I 9 : BRIDESMAID—Bridget of Troston 2041.

PROBATIONERS :—Baly 1st 1994.   Clarke 1st 2094.

SUMMARY :—Of the Elmham Group 1 Bull. Hudson Group 1 Cow. Probationers 2 Cows. Total 4.

THE RIGHT HON. LORD SUFFIELD, K.C.B.,

Gunton Park, Suffolk.   Post Town—Norwich.

Farm Bailiff—Mr. E. POLL.

A 1 : PRIMROSE—Primrose 1747.

A 2 : CHERRY—Daffodil 1425. Jessie 2268. Mabel 2343.

A 21 : ROSE—Bountiful 716. Rosa 1131.

A 22 : ALICE—Alice 671. Amy 674. Anna 1983. Laura 2303. Lilian 1622. Lily 999. Lovely 2329.

A 23 : DIDO—Dido 850. Freda 2199. Winifred 1273. Winsome 1924.

A 24 : FLOSS—Bridget 723. Fillpail 2187. Flora 1519. Florence 2195. Flower 901. Phœbe 2440.

A 25 : PANSY—Pansy 1718. Polly 2447. Poppy 1088. Venus 1247.

T 19 : TUCKER—Tucker 1966.

V 1 : COWSLIP—WILD RUFUS 778.

SUMMARY :—Of the Elmham Group 27 Cows. Walsingham Group 1 Cow. East Suffolk Group 1 Bull. Total 29.

MR. B. STIMPSON,

Morton-on-the-Hill, Norfolk.   Post Town—Norwich.

C 1 : BRISK—Zillah 2613.

D 3 : MARJORIE—Marjoram 2359. Miss Marjorie 2nd 1679.

HERDS AND OWNERS.

MR. STIMPSON'S HERD—*Continued.*

F 2 : BUTTERCUP—Beauty 692. Bountiful 2036.

F 4 : SNELLING—Satin 2525. Silky 1190. Superb 2561.

I 14 : JOY—Joy 955.

K 16 : NELLIE—Rosie 2nd 1811. Ruby 2517.

U 6 : PHŒNIX—Peter Piper 717.

W 3 : NELLY—Chapman 638. Charmer 2072. Chaste 2075. Cheerful 764. Cherry Pie 1385.

    Summary :—Of the Cranmer Group 1 Cow. Cley Group 2 Cows. Easton Group 5 Cows. Hudson Group 1 Cow. Kimberley Group 2 Cows. West Suffolk Group 1 Bull. Wolton Group 1 Bull 4 Cows. Total 17.

The Right Hon. the EARL OF STRADBROKE,

Henham Hall, Suffolk.

V 5 : CHERRY—Trimley True Heart 768.

H. LE STRANGE, Esq.,

Hunstanton Hall, Norfolk.   Post Town—King's Lynn.

Steward—Mr. John J. Newton.

A 6 : NORTON—Nightingale 2410. Redbreast 1766.

G 7 : FILLPAIL—Lapwing 987.

I 2 : RUBY—Ringdove 1782. Stockdove 2551.

I 3 : DAISY—Redpole 1768. Redshank 1773.

I 4 : HANDSOME—Linnet 1002. Redcrest 2486. Redstart 1777. Seagull 1839.

L 11 : LETTON—The Earl 664.

    Summary :—Of the Elmham Group 2 Cows. Hunstanton Group 1 Cow. Hudson Group 8 Cows. Dereham Group 1 Bull. Total 12.

HERDS AND OWNERS.

MR. G. K. TABER,

Pawling, New York, U.S.A.

AA 12 : HANDSOME—Maria 2357.

A 33 : ELM-LEAF—Celery 2061.

SUMMARY :—Of the American Group 1 Cow. Elmham Group 1 Cow.

MR. G. F. TABER,

Ravinewood Farm, Patterson, New York, U.S.A.

AA 1 : PRIMROSE—CHAMPION 271. Irene 2259. Lida 1620. Lucilla 1609. Lucretia 2385. Primrose Nell 2467. Ravinewood Lass 455.

AA 11 : NANCY—Tillie 1893.

AA 12 : HANDSOME—* Mabel 1649. May 1015. Melvina 1664. Mollie 1681.

A 13 : SPOT—Spinster 1861. Spotless 2545.

AA 29 : BELLE—Rachel 1121. Racine 2480. Ravinewood Belle 454. Rebecca 1762. Rosalie 1786.

AF 6 : CLARA—Claret 2nd 2093.

AK 17 : CHERRY—Miranda 2372. Princess May 1754. Thornham Prize 2572.

AN 7 : SKELTON—Sadie 1835. Susie 1220.

AU 5 : PRIMULA—Callie 1356. Cauliflower 3rd 82. Cecilia 2059. Christina 792. Christina 2nd 2089. Cinderella 1393.

N 4 : ROSE—MASON 698.

N 6 : TIT—Rose 1793. Titmouse 1902.

O 2 : QUEEN—Handsome Ruby 2239.

O 12 : BEAUTY—Sprite 1866.

O 13 : STRAWBERRY—Sophia 2542.

O 14 : CHERRY—Handsome Rose 2238.

* In the Register the number of Mabel is by misprint given as 1659. Her Tribal number should read AA 12.

HERDS AND OWNERS.

MR. TABER'S HERD—*Continued.*

U 6 : PHŒNIX—Daisy of Ravinewood 2122.

V 2 : RED STOCKINGS—Red Beauty 2nd 2484.

V 14 : HONESTY—Glemham Rose 2nd 2225.

W 14 : CLARA—Emmeline 2165.

X 3 : COSSET—Camelia 2nd 2055.

SUMMARY :—Of the Elmham Group 2 Cows. American Group 1 Bull 28 Cows. Necton Group 1 Bull 2 Cows. Oakley Group 4 Cows. West Suffolk Group 1 Cow. East Suffolk Group 2 Cows. Wolton Group 1 Cow. Trimley Group 1 Cow. Total 43.

MR. ALFRED TAYLOR,

Starston Place, Norfolk.   Post Town—Harleston.

Steward—Mr. BALE.

K 19 : ROSE—Buffet 2046. Bumptious 1345. Buxom 1355.

O 3 : COWSLIP—JUDGE 681. Cossett 2nd 2103. Cousin 2108. Rubbish 2516. Summer Rose 2559.

O 5 : NOSEGAY—* STARSTON PRINCE 758. Nosegay 2nd 394.

O 8 : MARY GREY—Jewel 281. Jewess 1597. Juice 2279.

R 2 : LOVELY—ARABI 618. JAWKINS 678. PASSION 714. REBEL 734. Fame 1505. Fashion 1501. Flirt 894. Nancy 368. Naughty 1697. Nettle 1049. Novice 2420. Sly 1192.

S 4 : HOLKHAM—Nonsense 2414. Novel 1056.

W 14 : CLARA—KELPIE 685.

SUMMARY :—Of the Kimberley Group 3 Cows. Oakley Group 2 Bulls 8 Cows. Starston Group 4 Bulls 8 Cows. Stoke Group 2 Cows. Wolton Group 1 Bull. Total 28.

* See Corrected Entry.

HERDS AND OWNERS.

MR. GARRETT TAYLOR,

Whitlingham. Postal Address—Trowse House, Norwich.

A 1 : PRIMROSE—Bloom 1325. Bloomer 1326. Daisy 1430. Minnie 1672. Peggy 1962. Princess 1964. Red Spot 2488. Spot 1862. White Spot 1934.

A 3 : BRIGHT—Bright Spark 2042.

A 9 : FANNY—Cato 468.

A 30 : CAROLINE—Laura 2304.

E 2 : CHERRY—Thirza 1889.

F 6 : POLLY—Moll 1680. Polly 1737. Polly 1082.

F 8 : FANNY—Fairy 881.

H 2 : BUTLER—Daisy Chain 2123. Red Daisy 2487.

H 4 : OLIVE—Olive 2nd 1714. Olly 2425.

K 15 : FILLPAIL—Flighty 1515. Flirt 893. Giddy Gal 2210.

K 17 : CHERRY—Cherry 4th 768.

M 2 : RED ROSE—Hester 267. Marham 2356.

N 23 : MASON—Maggie 1293.

N 24 : NECTON—Patsie 1724. Pattie 1297.

O 2 : QUEEN—Careless 1362.

O 9 : SILENCE—Dummy 2156.

O 11 : POLLY—Heartsease 2244. Pansy 1722.

P 1 : HANDSOME—Philip 538. Hannah 2240. Lydia 2342.

P 2 : STRAWBERRY—Isabel 1588. Ivy 2263.

P 3 : ROSE—Blush Rose 2032. Duchess 1468. Red Rose 1772. Rose 5th 1146. My Lady 2386.

P 10 : SALLY—Sally 526.

Q 1 : CHERRY—Icicle 1587. Nancy 1692.

MR. GARRETT TAYLOR'S HERD—*Continued.*

R 1 : COSSETT—Mysterious 2388. Mystery 2389. Mystic 2390. Sophia 2543.

R 8 : BEAUTY—Dingoo 2151. Dorcas 2153.

S 2 : HAPTON CHERRY—Hymen 1586.

S 3 : DOWSON—Quimbo 549. Damsel 1441. Damson 2124. Dido 2149. Dora 854.

T 1 : PRIMROSE—Countess 1408. Crown 1419. Primrose 2nd 440. Prune 1757.

V 9 : GLAD—Glaze 2212. Glow 1544.

V 11 : GLOSS—Press 2461.

V 13 : ROWLEY—Graceful 1546.

W 2 : BEAUTY—Water Fairy 1260. Winsome 1943.

W 14 : CLARA—Witch 1945.

PROBATIONERS :—Apple Girl 1987. Apple Sauce 1988. Appleton Brownie 1955. Cress 2112. Cresswell Brownie 1954.

SUMMARY :—Of the Elmham Group 1 Bull 11 Cows. Eaton Group 1 Cow. Easton Group 4 Cows. Hammond Group 4 Cows. Kimberley Group 4 Cows. Marham Group 2 Cows. Necton Group 3 Cows. Oakley Group 4 Cows. Powell Group 1 Bull 10 Cows. Stalham Group 2 Cows. Starston Group 6 Cows. Stoke Group 1 Bull 5 Cows. Thursford Group 4 Cows. East Suffolk Group 4 Cows. Wolton Group 3 Cows. Probationers 5 Cows. Total 75.

REV. CHARLES TERRY,

Harleston Hall, Stowmarket, Suffolk.

B 21 : ROSEBUD—SIR GARNET 752.

HERDS AND OWNERS.

CAPTAIN J. BORLASE TIBBITS,

Barton Seagrave, Northampton.   Post Town—Kettering.

Y 1 : BUTTERCUP—PLOUGHMAN 726. WAGGONER 776. Dark Beauty 2128. Darkie 3rd 842. Wallflower 1929. Wallflower 2nd 2603. Young Darkie 1949. Young Darkie 3rd 1950. Young Darkie 4th 1951. Young Gillyflower 2nd 1952.

Y 2 : CARNATION—Red Rose 1774. Rose 6th 1148. Rosemary 1818. White Rose 1933.

Y 3 : EVA—Buttercup 1350. Butterfly 4th 1351. Butterfly 5th 1352. Butterfly 7th 2051. Dolly 5th   . Dolly the Dairymaid 1467. Red-fly 1767.

SUMMARY :—Of the Barton Seagrove Group 2 Bulls 19 Cows. Total 21.

HENRY WARING, ESQ.,

Beenham House, Berks.   Post Town—Reading.

Steward—Mr. W. SALMOND.

A 31 : STAR—Strawberry 1870. Sybil 1224. Sybella 1225.

E 2 : CHERRY—Abbess 1967. Bessie 2nd 1321.

E 12 : SUSAN—Abigail 1969. Europa 1500.

F 3 : JOAN—Achievement 1975. Countess Beaconsfield 1411. Joanna 283.

F 8 : FANNY—Ballad 1993. Fuchsia 1528.

F 9 : BOO—Acacia 1970. Acceptance 1972. Brenda 1332. Minna 1671.

K 17 : CHERRY—CHARLES 469.

N 2 : MINNIE—Bessie 707.

P 4 : NINA—Accurate 1974. Alma 1981. Ethel 1498.

P 5 : RASPBERRY—Hyacinth 1584.

P 7 : VIOLET—Rosette 1821.

P 9 : CHERRY—Rosebud 1805.

HERDS AND OWNERS.

MR. HENRY WARING'S HERD—*Continued.*

W 4 : VIOLET—Accident 1973. Baubee 2000. Sphinx 1860.

SUMMARY :—Of the Elmham Group 3 Cows. Eaton Group 4 Cows. Easton Group 9 Cows. Kimberley Group 1 Bull. Necton Group 1 Cow. Powell Group 6 Cows. Wolton Group 3 Cows. Total 27.

MR. CHARLES WATERS,

Postwick Grange, Norfolk.   Post Town—Norwich.

A 1 : PRIMROSE—Nora 2415. Picket 2443. Pine 2445.
A 25 : PANSY—Pretty 2462.
E 12 : SUSAN—CHIEFTAIN 641.

SUMMARY :—Of the Elmham Group 4 Cows. Eaton Group 1 Bull. Total 5.

MR. WATSON,

Brampton Hall, Suffolk.

R 8 : BEAUTY—Rosy 2513.

MR. T. R. WEST,

Raveningham, Norfolk.   Post Town—Beccles.

R 9 : BRUNDISH—Polly 2449. Rosa 2492. Rosa 2nd 2493.

SUMMARY :—Of the Starston Group 3 Cows.

MR. HORACE WOLTON,

Newbourn Hall, Suffolk.   Post Town—Woodbridge.

W 1 : COSSETT—BATTERSEA BOURN 622. Battersea 4th 689. Battersea 5th 1998. Battersea Belle 1999.
W 2 : BEAUTY—PERFECT 536.

SUMMARY :—Of the Wolton Group 2 Bulls 3 Cows. Total 5.

MR. S. WOLTON,

Butley Abbey, Suffolk.   Post Town—Wickham Market.

W 1 : COSSETT—Battersea Boy 623.

W 2 : BEAUTY—Butley Boy 637. Beautiful Belle 2002. Belle of the Sands 2012. 4th Belle of Suffolk 43. 5th Belle of Suffolk 44. 6th Belle of Suffolk 704. 7th Belle of Suffolk 1316. 8th Belle of Suffolk 1317. 9th Belle of Suffolk 1318. 10th Belle of Suffolk 2013. 11th Belle of Suffolk 2014. 12th Belle of Suffolk 2015. Bridecake 2nd 1333. Bridesmaid 2nd 1339. Bridesmaid 3rd 2040. Butley Belle 2049.

W 3 : NELLY—Butley Beau 636. Betsy 2018. Ellen 1481. Nelly 3rd 377. Nelly 4th 1045. Star of the East 1868. Starry 2nd 1869. Venus 2nd 2591. Vesta 1248. Vesuvius 2595. Victoria 2598.

W 15 : DAISY—Tulip 3rd 2579.

Summary :—Of the Wolton Group 3 Bulls 26 Cows. Total 29.

CHANGES OF OWNERSHIP
to April 30th, 1883.

Victor 773 *from* Mr. R. H. Mason *to* the Rev. J. C. Girling.

Cellarette 2063
Rosinette 2511 } *from* Mr. R. H. Mason *to* Mr. G. Holmes.

Rupert   745
Countess 2106
Mabel   2344 } *from* Mr. N. Powell *to* Mr. T. Fulcher.
Maggie   2348
Minnie   2371

THE
RED POLLED RECORD,
For 1881-82.

HONOURS WON BY REGISTERED ANIMALS.

BULLS.

Alonso 447—2nd Norfolk, r. Suffolk, 1882.
Blount 457—2nd Norfolk, 1881.
Cortes 645—2nd Royal, 1st Norfolk, 1st Suffolk, and r. for cup Suffolk, 1882.
Davyson 3rd 48—1st Royal, 1st Norfolk, 1881. 1st Norfolk, 1st Suffolk, and cup, 1882.
Davyson 7th 476—1st Royal, 2nd Suffolk, 1881. 1st Royal, r. and h. c. Norfolk, 2nd Suffolk, 1882.
Davyson 10th 479—C. Norfolk, c. Suffolk, 1881.
Davyson 11th 480—H. c. Norfolk, 1882.
Favourite 492—C. Norfolk, 1881.
Frank 493—C. Norfolk, 1881.
The Friar 494—C. Suffolk, 1881.
Jumbo 683—C. Norfolk, 1882.
King Charles 329—2nd Royal, 2nd Norfolk, r. and h. c. Suffolk, 1882.
King Charming 687—R. and h. c. Royal, r. and h. c. Norfolk, r. and h. c. Suffolk, 1882.
King Tom 513—C. Norfolk, r. and h. c. Suffolk, 1881. 1st Suffolk, 1882.
Lord Bylaugh 690—1st Suffolk, 1881.
The Monk 525—2nd Norfolk, r. Royal, r. and h. c. Suffolk, 1881.
Othello 532—H., c. and r. Norfolk, 1st Wayland, 1881.
Passion 714—1st Royal, 1st Norfolk, 2nd Suffolk, 1882.
Pastor 715—2nd Norfolk, 1882.
Powerful 728—C. Norfolk, 1882.
Rinaldo 556—1st Royal, 1st Norfolk, 1st and cup Suffolk, 1881.
Rollick 558—H. c. Norfolk, 1881. C. Norfolk, h. c. Suffolk, 1882.
Roscoe 559—R. and h. c. Norfolk, 1882.
Rufus 188—2nd Royal, 2nd Norfolk, 1st and cup Suffolk, 1881.
Shylock 571—H. c. Suffolk, 1881. 1st Norfolk, r. and h. c. Suffolk, 1882.
Slasher 577—R. and h. c. Royal, c. Norfolk, 1882.
Starston Duke 570—2nd Royal, 2nd Norfolk, 1881. 2nd Norfolk, 1st Wayland, 1882.
Stout 581—H. c. Norfolk, 1881.
Wild Roger 603—C. Norfolk, 2nd Suffolk, 1881. R. and h. c. Norfolk, special prize and h. c. Suffolk, 1882.
Wild Rodney 602—1st Norfolk, h. c. Suffolk, 1881.
Wild Ruler 779—2nd Suffolk, 1882.

COWS.

Alice Brown—E 2. C. Norfolk, 2nd Suffolk, 1882.
B.B. 1315—T 6. C. Norfolk, 1881.
Beauty 2004—N 4. C. Norfolk, h. c. Wayland, 1882.
Bertha 1320—A 27. R. and h. c. Norfolk, 1882.
Blossom 1327—K 19. 2nd Royal, r. and h. c. Norfolk, 1881. 1st Norfolk and Norwich Christmas Show, 1882.
Buxom 1355—K 19. C. Norfolk, 2nd Suffolk, 1881.
Charmer 3rd 1368—K 18. H. c. Norfolk, 1881. C. Norfolk, 1882.
Cheerful 762—K 19. 2nd Wayland, 1882.
Cherry-leaf 1383—K 17. 1st Royal, 1st Norfolk, h. c. Suffolk, 1881. 1st Royal, 2nd Norfolk, 1st Suffolk, 1882.
Countess 1407—L 11. R. and h. c. Royal, 2nd Wayland, 1881.
Cousin 2108—O 3. C. Norfolk, 1882.
Daisy Girl 1439—O 7. C. Royal, c. Norfolk, 1881.
Davy 17th 846—H 1. C. Norfolk, 1881.
Davy 18th 847—H 1. 1st Norfolk, 1st Suffolk, 1st Wayland, 1881.
Davy 19th 848—H 1. C. Norfolk, 1881.
Davy 24th 1448—H 1. R. and h. c. Norfolk, c. Suffolk, 1881. C. Norfolk, h. c. Suffolk, 1882.
Davy 26th 1450—H 1. 2nd Norfolk, 1st Suffolk, 1st Wayland, 1881. C. Norfolk, 2nd Smithfield Club, 2nd Norfolk and Norwich Christmas Show, 1882.
Davy 37th 2130—H 1. 1st Royal, 1st Norfolk, 1882.

COWS—Continued.

Dolly 1463—N 2. 1st Royal, 1st Norfolk, 1st Suffolk, 1881, 1st and cup Norfolk, 2nd Suffolk, 1882.
Elmham 3rd 1485—A 3. C. Norfolk, 1882.
Empress 1496—N 4. R. Wayland, 1882.
Fame 1505—R 2. H. c. Norfolk, 1882.
Fancy 1598—V 3. 1st Norfolk, 1st and cup Suffolk, 1881.
Flirt 894—R 2. 1st Royal, 2nd Norfolk, 2nd Suffolk, 1881. 2nd Royal, 1st Norfolk, 2nd Suffolk, 1882.
Katie's Sister 1604—A 4. 1st Royal, r. and h. c. Norfolk, r. and h. c. Suffolk, 1882.
Melton Davy 1663—H 1. 2nd Norfolk, 1881. C. Norfolk, 1882.
Minnie 3rd 343—N 2. R. & c. Royal, 1881.
Minnie 6th 1674—N 2. C. Norfolk, 1881.
Minnie 7th 2368—N 2. H. c. Royal, c. Norfolk, 1882.
Moss Rose Queen 1685—A 2. C. Royal, 1881. C. Norfolk, r. & h. c. Suffolk, 1882.
Olivia 1715—H 4. 1st Smithfield Club, 1st Norfolk and Norwich Christmas Show, 1882.

Pet 1072—N 1. 1st Wayland, 1882.
Poppy 2456—U 43. C. Norfolk. 1881. 2nd Royal, 2nd Norfolk, 2nd Suffolk, 1st Wayland, 1882.
Rosamond 1789—P 3. 2nd Royal, 2nd Norfolk, 2nd Suffolk, 1881. 2nd Norfolk, 1st Suffolk, 1882.
Rosebud 1803—N 4. R. Wayland, 1882.
Satin 1837—T 7. C. Royal, c. Norfolk, 1881.
Silence 1853—O 9. 2nd Royal, h. c. Norfolk, r. and h. c. Suffolk, 1881.
Silent Lady 1855—O 9. 1st Norfolk, 1st Suffolk, 1882.
Tiny 1895—T 4. C. Norfolk, 1881.
Vermont Bridesmaid 2593—I 9. 2nd Royal, 2nd Norfolk, 1st Suffolk, 1882.
Vermont Topknot 2594—W 10. R. and h. c. Royal, r. and h. c. Norfolk, 1882.
Victoria 1926—N 6. R. and c. Norfolk, 1881. C. Norfolk, 1882.
Violet 4th 1252—N 6. 2nd Wayland, 1882.
Woodbine 1947—U 27. C. Norfolk, 1882.
Wild Cherry 2607—V 2. H. c. Norfolk, 1882.

CUP FOR BEST COLLECTION OF STOCK.

Norfolk, 1881, won by Mr. J. J. Colman, M.P.; r. and h. c. Mr. Jno. Hammond.
Norfolk, 1882, „ Mr. J. J. Colman, M.P.; 2nd Mr. A. Taylor; r. Mr. W. A. T. Amherst, M.P.
Suffolk, 1882, „ Mr. J. J. Colman, M.P.; 2nd Mr. W. A. T. Amherst, M.P.; r. Mr. A. Taylor.

PRIZE AS BEST MILKER.

Wild Rose 1271—V 1. Suffolk, 1881, 2nd Suffolk, 1882.
Wild Rose of Kilburn 1939—V 1. Essex, 1882.

PUBLIC SALES OF REGISTERED STOCK.

At the Suffolk Agricultural Society's Show, July 1st, 1881, Mr. Gooderham sold five young bulls, at an average of £19. 6s. 5d., and two cows at an average of 19 gs. Two yearling bulls from Mr. Collins' herd averaged 21 gs.

At Mr. Lofft's sixth annual sale, July 21st, 1881, the top price for bulls was 45 gs., for a bull of the N 2 tribe; for heifers 29 gs., for a calf of the U 43 tribe. Most of the animals offered were calves. The seventh sale, July 6th, 1882, was also for the most part an offer of calves; top price 43 gs., and 35 gs. for heifer calves for export to America. Two heifer calves were purchased by a Suffolk breeder at 25 gs. each. The top price for bull calves was 20½ gs.

Mr. T. Brown's annual sale, August 4th, 1881, of in-calf heifers, gave a top price of 35 gs. for one of the E 2 tribe. The average was £26. 7s. At the sale on August 3rd, 1882, the eight in-calf heifers offered averaged £41. 4s.

Mr. Fulcher held a draft sale of fourteen heifers, September 13th, 1881—average £14. 11s. 6d.

At Messrs. Spelman's first annual sale of registered stock, October 29th, 1881, the top price was 20½ gs. for a cow from Gunton. At the second sale, October 28th, 1882, eight in-calf heifers from Gunton, and several cows and bulls, were offered—top price 28½ gs. for Davyson 10th.

REGISTER OF BULLS.

613   ABBOT SAMPSON.
Calved October 8, 1881; breeder Mr. H. Biddell; s. St. Edmund 580; d. Honeysuckle 1574 by Iron Duke 125; gr d. June Rose 968 by The Baron 10; 2nd gr d. Rose—B 9.

614   ALDEBY.
Calved January, 1882; breeder Mr. T. Easter; owner Mr. Clarke; s. Brundish Prince 462; d. Nancy 2496 by Read 385; gr d. Brundish—R 9 by a Laxfield Bull.

615   ADONIS.
Calved April 3, 1881; breeder Mr. J. J. Colman, M.P.; owner the Right Hon. the Earl of Kimberley; s. Rufus 188; d. Beauty 1312 by Peter Powell 370; gr d. Blue Bell 52 by Norfolk Duke 127; 2nd gr d. Nina 2nd 389 by Tenant Farmer 213; 3rd gr d. Nina—P 4.

616   ALFRED.
Calved March 5, 1882; breeder Mr. G. F. Taber; owners Messrs. J. H. and W. W. Clark; s. Champion 271; d. Cauliflower 3rd 82 by Shylock 196; gr d. Cauliflower 81 by Sampson 191; 2nd gr d. Primula—U 5.

617   A LIVE BULL.
Calved December 22, 1881; breeder Mr. R. E. Lofft; s. Rinaldo 556; d. Wideawake 655 by Esquire 69; gr d. Starry 563 by Garibaldi 73; 2nd gr d. Nelly 375 by Robinson 178; 3rd gr d. Cherry—W 3 by Orwell 135.

REGISTER OF BULLS.

618 ARABI.
Calved February 18, 1882; breeder Mr. A. Taylor; s. Starston Duke 570; d. Flirt 894 by Easton Duke 61; gr d. Sly 1192 by Sir Edward I. 197; 2nd gr d. Strawberry 2nd 575 by Richard II. 173; 3rd gr d. Tiny—R 2 by Laxfield Sire 101.

619 ARABI.
Calved March 8, 1882; breeder Mrs. E. Perkins; s. Othello 532; d. Blossom 1328 by King Cole 330; gr d. Blossom by Prince Charlie 151; 2nd gr d. Violet 629 by Necton 3rd 122; 3rd gr d. Darling 2nd 162 by Necton 2nd 121; 4th gr d. Darling—N 1.

620 BAD TIMES.
Calved August, 1881; breeder Mr. G. Collins; s. Stout 581; d. Hunston Topknot 1579 by Oliver 364; gr d. Topknot 609 by Duke of Suffolk 57; 2nd gr d. Cherry—W 10.

621 BARON ROSCOE.
Calved December 3, 1882; breeder the Right Hon. Lord Hastings; s. Roscoe 559; d. Baroness Davy 1997 by Davyson 4th 286; gr d. Davy 19th 848 by Davyson 3rd 48; 2nd gr d. Davy 12th 174 by The Baron 9; 3rd gr d. Davy 5th 167 by Tenant Farmer 213; 4th gr d Davy—H 1.

622 BATTERSEA BOURN.
Calved December, 1882; breeder Mr. Horace Wolton; s. Perfect 536; d. Battersea 4th 689 by Oakley 133; gr d. Battersea 2nd 22 by Garibaldi 73; 2nd gr d. Heiress 264 by Battersea Eclipse 14; 3rd gr d. Battersea Favourite 21 by Robinson 178; 4th gr d. Cossett 119 by Rosehill 179; 5th gr d. Lovely—W 1.

623 BATTERSEA BOY.
Calved December 31, 1881; breeder Mr. H. Wolton; owner Mr. S. Wolton; s. May Prince 349; d. Battersea 4th 689 by Oakley 133; gr d. Battersea 2nd 22 by Garibaldi 73; 2nd gr d. Heiress 264 by Battersea Eclipse 14; 3rd gr d. Battersea Favourite 21 by Robinson 178; 4th gr d. Cossett 119 by Rosehill 179; 5th gr d. Lovely—W 1.

624 BAYARD.
Calved January 21, 1882; breeder Mr. T. Brown; s. Priam 373; d. The Elmham Belle 202 by Hero 2nd 86; gr d. Minnie—N 2 by Necton Prize 120.

625 BLUE BEARD.
Calved July 15, 1880; breeder Mr. H. Biddell; sire Ironsides 509; d. Picotee 410 by Great Britain 80; gr d. Picket—B 20.

626 BOLD HEART.
Calved January 17, 1882; breeder Mr. H. Biddell; owner Mr. A. J. Smith; s. Monarch 4th 351; d. White Heart 651 by Seneca 125; gr d. Cherry—V 5.

627 BON BON.
Calved December 24, 1881; breeder Mr. H. Birkbeck; s. Haman 499; d. Bonnie 53 by Norfolk Duke 127; gr d. Rose 2nd 479 by Tenant Farmer 213; 2nd gr d. Rose—P 3.

628 BONNET ROUGE.
Calved April 27, 1882; breeder Mr. W. A. T. Amherst, M.P.; s. Troston 3rd 591; d. Red Stockings 2nd 1128 by Councillor 38; gr d. Flora 2nd 897 by Doncaster 50; 2nd gr d. Flora 229 by King Alfred 96; 3rd gr d. Red Stockings—V 2 by Wonder 230.

629 BOWLER.
Calved May 4, 1882; breeder Mr. Fulcher; s. Lofty 515; d. Brilliant 728 by The Palmer 138; gr d. Brindy 729 by Hero 2nd 86; 2nd gr d. Curson—A 27.

630 BRIDEGROOM 2ND.
Calved September, 1877; breeder Mr. Lofft; s. Donald 291; d. Bridesmaid 2nd 721 by Rudham Hero 183; gr d. Bridesmaid—I 9.

631 BROWNLOW.
Calved May 17, 1882; breeder Mr. H. Biddell; owner the Right Hon. W. H. Smith; s. The Friar 494; d. Queen Rose 1759 by Monarch 4th 351; gr d. Rose—B 9.

632 BRUMMELL.
Calved June 28, 1881; breeder Mr. H. Birkbeck; owner Mr. F. J. Mann; s. Beau 259; d. Tit 3rd 607 by Norfolk Duke 127; gr d. Tit—T 4.

633   BRUTUS 3RD.

Calved July 15, 1881; breeder Mr. John Baly; s. Brutus Duo 463; d. Eleanor 1478 by Davyson 3rd 48; gr d. Helene 945 by Powell 143; 2nd gr d. Helen 266 by Norfolk Duke 127; 3rd gr d. Nelly of Newbourn 378 by Prince Regent 153; 4th gr d. Nelly 375 by Robinson 178; 5th gr d. Cherry—W 3.

634   BRUTUS 4TH.

Calved January 28, 1883; breeder Mr. John Baly; s. Brutus Duo 463; d. Lady Caroline 1610 by Roundhead 400; gr d. Lady-bird 977 by Major 109; 2nd gr d. Duchess of Suffolk—O 1.

635   BUSTLE.

Calved August 23, 1882; breeder Her Grace The Duchess of Hamilton; s. The Suffolk Baronet 583; d. Jessie 961 by Theodore 417; gr d. Jenny 279 by the Peer 139; 2nd gr d. Ruby 4th 518 by Tenant Farmer 213; 3rd gr d. Ruby 2nd 517 by Hero 2nd 86; 4th gr d. Ruby—I 2.

636   BUTLEY BEAU.

Calved August 20, 1882; breeder Mr. S. Wolton; s. Wild Rover 605; d. Star of the East 1868 by Perfection 2nd 368; gr d. Venus 623 by Duke of Suffolk 57; 2nd gr d. Starry 563 by Garibaldi 73; 3rd gr d. Nelly 375 by Robinson 178; 4th gr d. Cherry—W 3 by Orwell 135.

637   BUTLEY BOY.

Calved October 21, 1882; breeder Mr. Samuel Wolton; s. Wild Rover 605; d. 5th Belle of Suffolk 44 by Duke of Suffolk 58; 4th Belle of Suffolk 43 by Garibaldi 73; 2nd gr d. Topsy 611 by Robinson 178; 3rd gr d. Belle of Suffolk 41 by Orwell 135; 4th gr d. Beauty—W 2.

638   CHAPMAN.

Calved January 16, 1882; breeder Mr. B. Stimpson; s. Robin Hood 394; d. Cheerful 764 by Rufus 187; gr d. Cherry 2nd 101 by Cherry Duke 32; 2nd gr d. Cherry 100 by Duke of Suffolk 56; 3rd gr d. Nelly 375 by Robinson 178; 4th gr d. Cherry—W 3 by Orwell 135.

639   CHARLES.

Calved January, 1878; breeder Mr. T. Easter; s. Harold 83; d. Cherry—R 10 by a Laxfield Bull.

640   CHERRY PRINCE.

Calved January 6, 1882; breeder the Most Hon. the Marquis of Bristol; s. Fancy King 491; d. Double Cherry 855 by The Baron 10; gr d. Cherry—V 5.

641   CHIEFTAIN.

Calved March 22, 1882; breeder Mr. J. F. Rogers; owner Mr. Charles Waters; s. Emperor 489; d. Endive 1497 by Osman 530; gr d. Susanna 587 by Stoke Duke 209; 2nd gr d. Susan—E 12.

642   COLONEL, THE.

Calved November 14, 1882; breeder Mrs. E. Perkins; s. Othello 532; d. Daisy 3rd 1434 by Bradfield 264; gr d. Daisy 152 by Necton 3rd 122; 2nd gr d. Rose—N 4 by Necton Prize 120.

643   COMMANDER.

Calved March 24, 1882; breeder Mr. G. F. Taber; owner Messrs. D. and G. Jones; s. Champion 271; d. Callie 1356—AU 5 by Ravinewood Beau 160; gr d. Cauliflower 3rd 82 by Shylock 196; 2nd gr d. Cauliflower 81 by Sampson 191; 3rd gr d. Primula—U 5.

644   COPFORD PRINCE.

Calved Dec. 22, 1880; breeder Mr. A. Taylor; owner Mr. T. H. Harrison; s. King Charles 329; d. Lovely 2nd 322 by Richard II. 173; gr d. Pretty 425 by Richard I. 172; 2nd gr d. Lily 313 by Laxfield Sire 101; 3rd gr d. Lovely—R 2 by Laxfield Sire 101.

645   CORTES.

Calved August 18, 1881; breeder Mr. R. E. Lofft; owner Mr. W. A. T. Amherst, M.P.; s. Stout 581; d. Handsome 8th 1554 by Bright 267; gr d. Handsome 5th 935 by Troston Hero 221; 2nd gr d. Handsome 2nd 249 by Sampson 191; 3rd gr d. Handsome—U 3.

646   COUNT DAVY.

Calved January 6, 1883; breeder Mr. W. A. T. Amherst, M.P., s. Davyson 3rd 48; d. Waxwork 6th 1932 by Hector 319; gr d. Waxwork 2nd 648 by King of Carlford 100; 2nd gr d. Waxwork—U 9.

647   CROMWELL.

Calved November 10, 1882; breeder Mr. J. J. Colman, M.P., s. Roundhead 564; d. Dolly 1463 by Rufus 188; gr d. Polly 1084 by Rufus 189; 2nd gr d. Lily 2nd 311 by Hero 3rd 87; 3rd gr d. Lily 310 by Hero of Newcastle 85; 4th gr d. Minnie—N 2 by Necton Prize 120.

648   CYPRUS 2ND.

Calved November 21, 1882; breeder the Right Hon. Lord Henniker; s. Cyprus 473; d. Polly 1085 by Eclipse 2nd 299; gr d. Thornham Polly 1229 by Eclipse 63; 2nd gr d. Polly—O 11.

649   DALE.

Calved January 27, 1882; breeder Mr. R. E. Lofft; s. Long 516; d. Elmham 2nd 1484 by Bright 267; gr d. Elmham 199 by Hero 3rd 87; 2nd gr d. Brettenham Handsome—A 3 by Hero of Newcastle 85.

650   DALLINGHOO.

Calved July 27, 1881; breeder Mr. G. Gooderham; s. Watchman 777; d. Flirt 2nd 1516 by Troston 424; gr d. Flirt 895 by Councillor 38; 2nd gr d. Rosebud 497 by Doncaster 50; 3rd gr d. Rosy 513 by Wonder 230; 4th gr d. Cowslip—V 1.

482   DAVYSON 13TH.
[Corrected Entry.]

Calved November, 1880; breeder Mr J. Hammond; owner Mr. E. Cooke; s. Norfolk John 2nd 527; d. Davy 24th 1448 by Davyson 5th 287; gr d. Davy 15th 844 by Davyson 3rd 48; 2nd gr d. Davy 5th 167 by Tenant Farmer 213; 3rd gr d. Davy—H 1.

651   DAVYSON 14TH.

Calved February, 1881; breeder Mr. John Hammond; owner Mr. J. Margarson; s. Davyson 7th 476; d. Davy 5th 167 by Tenant Farmer 213; gr d. Davy—H 1.

652   DAVYSON 15TH.

Calved September 22, 1881; breeder Mr. J. Hammond; s. Davyson 6th 475; d. Davy 10th 172 by Sir Nicholas 2nd 203; gr d. Davy 3rd 165 by Sir Nicholas 202; 2nd gr d. Rose of Hope 506 by Hammond's Rufus 82; 3rd gr d. Davy—H 1.

653 DAVYSON 16TH.

Calved September 1881; breeder Mr. J. Hammond; s. Davyson 7th 476; d. Davy 22nd 1446 by Davyson 5th 287; gr d. Davy 16th 845 by Redjacket 7th 169; 2nd gr d. Davy 7th 169 by Young Duke 234; 3rd gr d. Davy 2nd 164 by Sir Nicholas 202; 4th gr d. Davy—H 1.

654 DEXTER.

Calved March 30, 1881; breeder Mr. G. F. Taber; owner General L. F. Ross; s. Ravinewood Beau 160; d. Ravinewood Belle 454 by Hero 3rd 87; gr d. Mrs. Rollin—A 29.

655 DIAMOND.

Calved September 22, 1880; breeder Mr. J. J. Colman, M.P.; s. Rufus 188; d. Polly 1084 by Rufus 189; gr d. Lily 2nd 311 by Hero 3rd 87; 2nd gr d. Lily 310 by Hero of Newcastle 85; 3rd gr d. Minnie—N 2 by Necton Prize 120.

656 DIDLINGTON DAVYSON.

Calved January 6, 1883; breeder Mr. W. A. T. Amherst, M.P.; s. Davyson 3rd 48; d. Davy 30th 1454 by Davyson 6th 475; gr d. Davy 22nd 1446 by Davyson 5th 287; 2nd gr d. Davy 16th 845 by Redjacket 7th 169; 3rd gr d. Davy 7th 169 by Young Duke 234; 4th gr d. Davy 2nd 164 by Sir Nicholas 202; 5th gr d. Davy—H 1.

657 DIDLINGTON DAVYSON 2ND.

Calved January 7, 1883; breeder Mr. W. A. T. Amherst, M.P.; s. Davyson 12th 481; Davy 24th 1448 by Davyson 5th 287; gr d. Davy 15th 844 by Davyson 3rd 48; 2nd gr d. Davy 5th 167 by Tenant Farmer 213; 3rd gr d. Davy—H 1.

658 DOCTOR DAVYSON.

Calved January 18, 1882; breeder Mr. John Howell; s. Davyson 9th 478; d. Nancy 1039 by Masker 346; gr. d. The Nun—T 10.

659 DON CARLOS.

Calved November 18, 1882; breeder Mr. J. J. Colman, M.P.; s. King Charles 329; d. Miss Atkins 1023 by Powell 143; gr d. Lady Atkins 290 by Norfolk Duke 127; 2nd gr d. Primrose 433 by Tenant Farmer 213; 3rd gr d. Cherry—K 17.

660   DON PEDRO.

Calved March 1, 1881; breeder Mr. Garrett Taylor; owner Mr. J. M. Spinks; s. Grey Spot 498; d. Isabel 1588 by Duke of Norfolk 295; gr d. Isabelle 278 by Young Duke 234; 2nd gr d. Strawberry 2nd 573 by Tenant Farmer 213; 3rd gr d. Strawberry—P 2.

661   DONCASTER.

Calved July 5, 1882; breeder Her Grace the Duchess of Hamilton; s. The Wilby Lad 599; d. Radish 1761 by Norfolk Duke 127; gr d. Rosy 510 by Duke 52; 2nd gr d. Rose of Eaton 504 by Cringleford Sire 44; 3rd gr d. Cowslip 125 by Stoke 208; 4th gr d. Rose 470 by Son of Hapton 205; 5th gr d. Cherry—E 2.

662   DRUID.

Calved April 29, 1882; breeder Mr. Garrett Taylor; owner Mr. J. J. Colman, M.P.; s. Davyson 8th 477; d. Water Fairy 1260 by Umpire 223; gr d. Whitethorn 654 by Powell 143; 2nd gr d. Topsy 611 by Robinson 178; 3rd gr d. Belle of Suffolk 41 by Orwell 135; 4th gr d. Beauty—W 2.

663   DUKE OF DAYTON.

Calved April 17, 1882; breeder Mr. G. F. Taber; owner Mr. J. M. Lain Smith; s. Champion 271; d. Rebecca 1762—AA 29 by Ravinewood Beau 160; gr d. Ravinewood Belle 454 by Hero 3rd 87; 2nd gr d. Mrs. Rollin—A 29.

664   THE EARL.

Calved January 23, 1882; breeder Mr. W. A. T. Amherst, M.P.; owner Mr. le Strange; s. Davyson 3rd 48; d. Countess 1407 by Roger 396; gr. d. Lass 988 by an Elmham Bull; 2nd gr d. Letton—L 11.

665   EARL DAVY.

Calved Dec. 17, 1882; breeder Mr. W. A. T. Amherst, M.P.; s. Davyson 3rd 48; d. Gentle Rose 914 by Iron Duke 125; gr d. Dwarf Rose 193 by Cremorne 42; 2nd gr d. Rose—B 9.

666   EATON.

Calved May 11, 1881; breeder Mr. J. F. Rogers; owner Rev. J. C. Girling; s. Robin Hood 394; d. Esther 874 by Nicholson 360; gr d. Eaton Beryl 804 by Powell 143; 2nd gr d. Cherry 86 by Stoke 208; 3rd gr d. Countess—E 3.

REGISTER OF BULLS.

667 FABIAN.

Calved December 16, 1882; breeder Mr. T. Brown; s. Priam 373; d. Faith 1504 by Norfolk Duke 127; gr d. Florence 898 by The Palmer 38; 2nd gr d. Thursford Rose 600 by Norfolk Duke 127; 3rd gr d. Rose—P 3.

668 FESTUS.

Calved November 13, 1882; breeder Mr. J. F. Rogers; s. Emperor 489; d. Fanny—F 8.

669 FRANCILLO.

Calved January 13, 1882; breeder Mr. J. J. Colman, M.P.; s. Charles 469; d. Fan 1506 by Roundhead 180; gr d. Fanny 220 by Hero 3rd 87; 87; 2nd gr d. Madame Freeman—A 9.

670 FRANK OF ELMHAM.

Calved November 24, 1881; breeder Rev. A. G. Legge; owner Mr. Fulcher; d. Lofty 515; d. Frances 904 by Harry 84; gr d. Fanny Bradfield 891 by Money 352; 2nd gr d. Nancy—A 11.

671 FRITZ.

Calved December 18, 1882; breeder the Right Hon. Lord Hastings; s. Roscoe 559; d. Fillpail 1512 by Davyson 3rd 48; gr d. Nancy 1690 by Peck 534; 2nd gr d. Spot 3rd 1863 by Wilby Chapman 228; 3rd gr d. Spot 558 by Wonder 231; 4th gr d. Rose—K 19 by an Elmham Bull.

672 FUSILIER.

Calved November 12, 1882; breeder Mr. T. Brown; s. Priam 373; d. Fusee 1530 by Royal Duke 181; gr d. Florence 898 by The Palmer 38; 2nd gr d. Thursford Rose 600 by Norfolk Duke 127; 3rd gr d. Rose—P 3.

673 GLANDFORD PRINCE.

Calved November 28, 1881; breeder Mr. N. Powell; s. Norfolk John 2nd 527; d. Primrose 3rd 1749 by Norfolk John 131; gr d. Polly 416 by a Red Polled Bull; 2nd gr d. Violet—P 7.

674 GRASSHOPPER.

Calved August 17, 1881; breeder Mr. H. Birkbeck; owner Mr. P. Leeder; s. Haman 499; d. Elmham Georgiana 1487 by Rufus 188; gr d. Pansy 1063 by Cringleford Duke 43; 2nd gr d. Pretty 422 by Cantley 29; 3rd gr d. Polly—E 11 by Duke 52.

675 HECTOR.
Calved December 3, 1882; breeder Mr. N. Powell; s. Premier 543; d. Hetty 1570 by Norfolk John 2nd 527; gr d. Holly 1571 by Norfolk John 131; 2nd gr d. Handsome 3rd by Norfolk John 131; 3rd gr d. Handsome 2nd by Norfolk Duke 127; 4th gr d. Raspberry—P 5.

676 HIGH SHERIFF.
Calved October 10, 1880; breeder Mr. W. G. Collins; s. Ross 562; d. Viscountess 2nd 1928 by Prince 376; gr d. Viscountess—U 48 by Prince Arthur 150.

677 HUNSTON DUKE 3RD.
Calved November 20, 1880; breeder Mr. W. G. Collins; owner Mr. E. Boon's Executors; s. Tommy 587; d. Constance 2nd 800 by Cherry Duke 32; gr d. Constance 799 by Plowman 371; 2nd gr d. Weasel by Newcastle Prize 359; 3rd gr d. Cherry by Lord Manners 341; 4th gr d. Fancy—U 45 by Peter 369.

678 JAWKINS.
Calved May 3, 1882; breeder Mr. A. Taylor; s. Starston Duke 570; d. Nettle 1049 by Easton Duke 61; gr d. Nancy 363 by Richard II. 173; 2nd gr d. Lovely 2nd 322 by Richard II. 173; 3rd gr d. Pretty 425 by Richard I. 172; 4th gr d. Lily 313 by Laxfield Sire 101; 5th gr d. Lovely—R 2 by Laxfield Sire 101.

679 JOSEPHUS.
Calved March 27, 1882; breeder Mr. G. F. Taber; owner Mr. S. B. Douglas; s. Champion 271; d. Ravinewood Belle 454 by Hero 3rd 87; gr d. Mrs. Rollin—A 29.

680 JUDGE.
Calved September 4, 1882; breeder Mr. G. F. Taber; owner Mr. J. L. Mustard; s. Red Knight 735; d. Sprite 1866 by Crown Prince 281; gr d. Sprightly 2nd 1201 by Eclipse 2nd 299; 2nd gr d. Sprightly 1200 by Eclipse 63; 3rd gr d. Beauty—O 12.

681 JUDGE.
Calved August 11, 1882; breeder Mr. A. Taylor; s. Starston Duke 570; d. Cossett 2nd 2103 by King Charles 329; gr d. Cossett 1405 by Rifleman 175; 2nd gr d. Cowslip—O 3 by Bowbearer 22; 3rd gr d. by a Glemham Bull.

682   JULIAN.

Calved December 1, 1882; breeder Mr. T. Brown; s. Priam 373; d. Juliet 967 by Norfolk Duke 127; gr d. Jenny 279 by The Peer 139; 2nd gr d. Ruby 4th 518 by Tenant Farmer 213; 3rd gr d. Ruby 2nd 517 by Hero 2nd 86; 4th gr d. Ruby—I 2.

683   JUMBO.

Calved October 7, 1881; breeder Her Grace the Duchess of Hamilton; s. The Suffolk Baronet 583; d. Little Katie 1630 by Royal Duke 181; gr d. Kattie 975 by Benedict 17; 2nd gr d. Ringlet 2nd 465 by Tenant Farmer 213; 3rd gr d. Ringlet 464 by Hero of Newcastle 85; 4th gr d. Brettenham Strawberry—A 4 by Redjacket 163.

684   JUMBO.

Calved February 10, 1882; breeder Mr. C. Austin; s. Shylock 571; d. Sweetbloom 1221 by The Baron 10; gr d. Rosebloom 485 by Seneca 195; 2nd gr d. Rosebud—B 21.

685   KELPIE.

Calved March 28, 1881; breeder Mr. Garrett Taylor; owner Mr. A. Taylor; s. Grey Spot 498; d. Water Witch 1264 by Norfolk Duke 127; gr d. Witch 657 by Tommy 216; 2nd gr d. Clara—W 14.

686   KING BRAMBLE.

Calved October 19, 1881; breeder Mr. H. Biddell; owner Mr. Bantoft, jun.; s. St. Edmund 580; d. Blackberry Jam 1324 by Crown Prince 281; gr d. Cherry Jam 786 by The Baron 10; 2nd gr d. Cherry Lux—B 2.

687   KING CHARMING.

Calved September 22, 1881; breeder Mr. J. J. Colman, M.P.; s. Rufus 188; d. Rosebud 2nd 1797 by Rufus 188; gr d. Rosebud 494 by Norfolk Duke 127; 2nd gr d. Cherry 2nd 91 by Tenant Farmer 213; 3rd gr d. Cherry—K 17.

688   KING EGBERT.

Calved February 10, 1882; breeder Mr. T. Fulcher; owner Mr. W. A. T. Amherst, M.P.; s. Lofty 515; d. Spotless 1197 by The Palmer 138; gr d. Twinny 620 by Hero 3rd 87; 2nd gr d. Nancy—A 11.

689   LANCER.

Calved November 1881; breeder Mr. W. Bradfield; s. Tommy 588; d. Lettice 2309 by Falstaff 303; gr d. Lucy 2338 by an Elmham Bull; 2nd gr d. a Pond Cow—1 Norf.

612   LONGFELLOW.

Calved July 4, 1880; breeder Mr. Garrett Taylor; (exported to New Brunswick;) s. Grey Spot 104; d. Minnie—N 23 by Fransham Captain 71.

690   LORD BYLAUGH.

Calved September 6, 1880; breeder Mr. Lofft; s. Ross 562; d. Minnie 6th 1674 by Hector 319; gr d. Minnie 4th 1022 by Robin Hood 177; 2nd gr d. Minnie 3rd 343 by Hammond 81; 3rd gr d. Minnie—N 2 by Necton Prize 120.

691   LORD BYLAUGH 2ND.

Calved January 26, 1882; breeder Rev. H. Evans Lombe; s. Lord Elmham 519; d. Joyful 1598 by Othello 532; gr d. Nancy 2nd 1691 by Young Major 235; 2nd gr d. Nancy 1690 by Peck 534; 3rd gr d. Spot 3rd 1863 by Wilby Chapman 228; 4th gr d. Spot 558 by Wonder 231; 5th gr d. Rose—K 19 by an Elmham Bull.

692   LORD BYLAUGH 3RD.

Calved March 5, 1882; breeder Rev. H. Evans Lombe; s. Lord Elmham 519; d. Blossom 2024 by a Ramsley Bull; gr d. Ramsley—A 5 by Hero of Newcastle 85.

693   LORD CHARLES.

Calved July 17, 1880; breeder Mr. R. E. Lofft; s. Slasher 577; d. Rosebud 3rd 1798 by Donald 291; gr d. Rosebud—I 13.

694   LORD DAVY.

Calved November 13, 1882; breeder Mr. W. A. T. Amherst, M.P.; s. Davyson 3rd 48; d. Flirt 3rd 1517 by Troston 424; gr d. Flirt 895 by Councillor 38; 2nd gr d. Rosebud 497 by Doncaster 50; 3rd gr d. Rosy 513 by Perfection 140; 4th gr d. Beauty 36 by Wonder 230; 5th gr d. Cowslip—V 1.

REGISTER OF BULLS.

695 LORD HARTISMERE.
Calved March 8, 1882; breeder The Right Hon. Lord Henniker; s. Cyprus 473; d. Sprightly 2nd 1201 by Eclipse 2nd 299; gr d. Sprightly 1200 by Eclipse 63; 2nd gr d. Beauty—O 12.

696 LUCAS.
Calved November 1882; breeder Mr. E. Boon's Executors; s. Wild Roger 603; d. Lovely 2330 by a Thornham Bull; gr d. Nancy—2 SUFF. by a Kettleburgh (Turner's stock) Bull.

697 MADCAP.
Calved October 17, 1882; breeder Her Grace the Duchess of Hamilton; s. The Suffolk Baronet 583; d. Theodosia 1227 by The Beau 16; gr d. Tulip 613 by Duke 52; 2nd gr d. Cowslip 2nd 126 by Spot 206; 3rd gr d. Cowslip 125 by Stoke 208; 4th gr d. Rose 470 by Son of Hapton 205; 5th gr d. Cherry—E 2.

698 MASON.
Calved May 10, 1881; breeder Mr. R. H. Mason; owner Mr. G. F. Taber; s. Slasher 577; d. Empress 1496 by King Harry 1332; gr d. Rose 2nd 1143 by Longham 104; 2nd gr d. Daisy 152 by Necton 3rd 122; 3rd gr d. Rose—N 4 by Necton Prize 120.

699 MELTON DAVID.
Calved December 5, 1882; breeder the Right Hon. Lord Hastings; s. Roscoe 559; d. Melton Davy 1663 by Thornham Duke 2nd 585; gr d. Davy 12th 174 by The Baron 9; 2nd gr d. Davy 5th 167 by Tenant Farmer 213; 3rd gr d. Davy—H 1.

700 MILITIAMAN.
Calved December, 1880; breeder Mr. J. W. Vincent; owner Mr. T. Fulcher; s. Premier 372; d. a Pond Cow bred by Wiffen—1 NORF.

701 MISCHIEVOUS.
Calved December 15, 1882; breeder Mr. John Baker; s. Brundish Prince 462; d. Nancy 2394 by Read 386; gr d. Brundish—R 9 by a Laxfield Bull.

702 NECTON DUKE.

Calved March 28, 1882; breeder Mr. W. A. T. Amherst, M.P.; s. Slasher 577; d. Dolly 852 by Lord Easton 105; gr d. Nina 1052 by Bradfield 264; 2nd gr d. Nancy 359 by Fransham Captain 71; 3rd gr d. Tit—N 6 by Necton 3rd 122.

703 NEGRO.

Calved November 24, 1881; breeder Mr. Lofft; owner Mr. Robt. Ives; s. Rinaldo 556; d. Newbourn Pride 8th 1709 by Duke 488; gr d. Newbourn Pride 4th 1051 by Cherry Duke 32; 2nd gr d. Newbourn Pride 2nd 384 by Glatton 79; 3rd gr d. Newbourn Pride 383 by Garibaldi 73; 4th gr d. Nelly 375 by Robinson 178; 5th gr d. Cherry—W 3 by Orwell 135.

704 NIPPENOSE BEAU.

Calved January 4, 1881; breeder Mr. G. F. Taber; owner Mr. G. L. Sanderson; s. Champion 271; d. Susie 1220—AN 7 by Ravinewood Beau 160; gr d. Skelton—N 7 by Necton 3rd 122.

705 NIPPER.

Calved May, 1882; breeder Mr. W. Bradfield; owner Mr. Jas. Rivett; s. Tommy 588; d. Nellie 1702 by The Palmer 138; gr d. Nelly 371 by Hero 2nd 86; 2nd gr d. Primrose—A 1 by Elmham Sire 67.

706 NOBBY.

Calved June 16, 1882; breeder Mr. Fulcher; s. Lofty 515; d. Nancy 1689 by Rufus 188; gr d. Fenn—A 37 by The Palmer 138; 2nd gr d. by Hero 3rd 86.

707 NO DOUBT.

Calved March 12, 1882; breeder Mr. R. E. Lofft; s. Doubtful 487; d. Rosebud 5th 1800 by Prince 377; gr d. Rosebud 2nd 1153 by Rudham Hero 183; 2nd gr d. Rosebud—I 13.

708 NONSUCH TOM.

Calved November 4, 1882; breeder Mr. C. K. Cordy; s. Trimley Tom 589; d. Trimley Nonsuch 1912 by a Son of Rifleman 175; gr d. Minnie 1021 by Harold 83; 2nd gr d. Mary Grey—O 8.

709 NORFOLK WIZARD.

Calved January 21, 1883; breeder Mr. W. A. T. Amherst, M.P.; s. Davyson 3rd 48; d. Norfolk Witch 1054 by Norfolk Duke 127; gr d. Witch 657 by Tommy 215; 2nd gr d. Clara—W 14.

710 NOT DOUBTFUL.

Calved November 26, 1881; breeder Mr. R. E. Lofft; owner Mr. C. K. Cordy; s. Doubtful 487; d. Waxwork 2nd 648 by King of Carlford 100; gr d. Waxwork—U 9.

711 ORLANDO.

Calved April 7, 1881; breeder Mr. H. Biddell; s. Monarch 4th 351; d. Carnation 745 by Crown Prince 281; gr d. Picotee 410 by Great Britain 80; 2nd gr d. Picket—B 20.

712 ORLANDO.

Calved October 20, 1881; breeder Mr. R. E. Lofft; owner Mr. R. H. Mason; s. Ross 562; d. Bridesmaid 6th 1336 by Donald 291; gr d. Bridesmaid 3rd 722 by Cherry Duke 32; 2nd gr d. Bridesmaid—I 9.

713 OTHELLO.

Calved December 1, 1881; breeder Mr. J. J. Colman, M.P.; s. Rufus 188; d. Miss Atkins 1023 by Powell 143; gr d. Lady Atkins 290 by Norfolk Duke 127; 2nd gr d. Primrose 483 by Tenant Farmer 213; 3rd gr d. Cherry—K 17.

714 PASSION.

Calved January 2, 1881; breeder Mr. A. Taylor; s. King Charles 329; d. Sly 1192 by Sir Edward I. 197; gr d. Strawberry 2nd 575 by Richard II. 173; 2nd gr d. Tiny 604 by Laxfield Sire 101; 3rd gr d. Lovely—R 2.

715 PASTOR.

Calved November 14, 1881; breeder Mr. A. Taylor; owner Sir E. C. Kerrison, Bart.; s. Starston Duke 570; d. Fashion 1510 by King Charles 329; gr d. Sly 1192 by Sir Edward I. 197; 2nd gr d. Strawberry 2nd 575 by Richard II. 173; 3rd gr d. Tiny 604 by Laxfield Sire 101; 4th gr d. Lovely—R 2 by Laxfield Sire 101.

716   PEPPERCORN.

Calved January, 1883; breeder Mr. P. Blofield; s. Rollick 558; d. Prune 1116 by Young Major 235; gr d. Princess by a Stoke Bull; 2nd gr d. Bride—K 25.

717   PETER PIPER.

Calved September 19, 1881; breeder Mr. Lofft; owner Mr. B. Stimpson; s. Stout 581; d. Phœnix 2nd 2442 by Hector 319; gr d. Phœnix—U 6.

718   PETITIONER.

Calved January 22, 1882; breeder Mr. R. H. Mason; s. Slasher 577; d. Pet 1072 by Lord Easton 105; gr d. Polly 414 by Fransham Captain 71; 2nd gr d. Darling 2nd 162 by Necton 2nd 121; 3rd gr d. Darling—N 1 by Necton Prize 120.

719   PETRARCH.

Calved July 30, 1882; breeder Mr. R. H. Mason; s. Philip 538; d. Phœbe 1733 by King Cole 330; gr d. Pet 1072 by Lord Easton 105; 2nd gr d. Polly 414 by Fransham Captain 71; 3rd gr d. Darling 2nd 162 by Necton 2nd 121; 4th gr d. Darling—N 1 by Necton Prize 120.

720   PICKWICK.

Calved 1877; breeder Her Grace the Duchess of Hamilton; owner Mr. A. J. Smith; s. Baron Handsome 254; d. Glemham Rose 921 by Young Monarch 246; gr d. Fillpail by Duke 239; 2nd gr d. Honesty—V 14.

721   PLATO.

Calved March 14, 1881; breeder Mr. T. Brown; s. Favourite 492; d. Penelope 1069 by Roundhead 180; gr d. Nelly 372 by Redjacket 7th 169; 2nd gr d. Handsome 2nd 244 by Tenant Farmer 213; 3rd gr d. Handsome—P 1.

722   PLAYFORD DUKE.

Calved August 9, 1882; breeder Mr. H. Biddell; s. Bluebeard 624; d. Piquet 1076 by The Baron 10; gr d. Picket—B 20.

723   PLAYMATE.

Calved December 21, 1880; breeder Mr. T. Brown; s. Favourite 492; d. Nelly 372 by Redjacket 7th 169; gr d. Handsome 2nd 244 by Tenant Farmer 213; 2nd gr d. Handsome—P 1.

724 PLINY.

Calved July 26, 1881; breeder Mr. T. Brown; s. Bergamot 455; d. Pensive 1729 by Norfolk Duke 127; gr d. Penelope 1069 by Roundhead 180; 2nd gr d. Nelly 372 by Redjacket 7th 169; 3rd gr d. Handsome 2nd 244 by Tenant Farmer 213; 4th gr d. Handsome—P 1.

725 PLOUGHBOY.

Calved March 10, 1882; breeder Mr. N. Powell; s. Premier 543; d. Priscilla 2nd 1965 by Norfolk John 2nd 527; gr d. Priscilla 1114 by Norfolk John 131; 2nd gr d. Nancy 2nd 361 by Norfolk Duke 127; 3rd gr d. Nancy—P 6.

726 PLOUGHMAN.

Calved February 11, 1883; breeder Capt. J. Borlase Tibbits; s. Plowboy 540; d. Young Darkie 1949 by Young Foxhall 437; gr d. Darkie 841 by a Suffolk Bull; 2nd gr d. Rose by a Suffolk Bull; 3rd gr d. Gillyflower by a Suffolk Bull; 4th gr d. Buttercup—Y 1.

727 POPE, THE

Calved March, 1881; breeder Her Grace the Duchess of Hamilton; owner Mr. H. Haylock; s. Handsome Prince 317; d. Rosie 1812 by Marquis 344; gr d. Glemham Rose 921 by Young Monarch 246; 2nd gr d Fillpail by Duke 239; 3rd gr d. Honesty—V 14.

728 POWERFUL.

Calved September 4, 1880; breeder Mr. R. E. Lofft; s. Hector 319; d. Poppet 3rd 1742 by Honest Tom 88; gr d. Poppet—U 43 by Sampson 191.

729 PRINCE ALBERT.

Calved January 18, 1881; breeder Mr. G. F. Taber; owner Mr. J. M. Knapp; s. Champion 271; d. Rebecca 1762—AA 29 by Ravinewood Beau 160; gr d. Ravinewood Belle 454 by Hero 3rd 87; 2nd gr d. Mrs. Rollin—A 29.

730 PRINCE CHARLIE.

Calved June 28, 1881; breeder Mr. H. Biddell; s. Monarch 4th 351; d. Pretty Flower 1093 by Iron Duke 125; gr d. Fancy Flower 219 by Seneca 195; 2nd gr d. Fancy—B 18.

731   PRINCE IMPERIAL.

Calved April 11, 1882; breeder Mr. R. H. Mason; s. Slasher 577; d. Empress 1496 by King Harry 332; gr d. Rose 2nd 1143 by Longham 104; 2nd gr d. Daisy 824 by Necton 3rd 122; 3rd gr d. Rose—N 4 by Necton Prize 120.

732   PROSPERO.

Calved November 29, 1882; breeder Mr. T. Leonard Palmer; s. Rollick 558; d. Lily 4th 1627 by Stout 581; gr d. Lily 3rd 1000 by The Palmer 138; 2nd gr d. Lily 310 by Hero of Newcastle 85; 3rd gr d. Minnie—N 2 by Necton Prize 120.

609   PRYOR.

Calved May 4, 1876; breeder Mr. Lofft; s. The Palmer 138; d. Lily by Hero of Newcastle 85; gr d. Minnie—N 2 by Necton Prize 120.

610   PUNCH.

Calved February, 1876; breeder Mr. W. Hudson; s. Quarles Duke 548; d. Margaret—I 19 by Quarles Duke 548; gr d. a Hudson Cow by Proud 547.

611   PURL.

Calved December 23, 1876; breeder Mr. J. Margarson; s. Master Freeman 347; d. Cherry—L 9.

733   RADICAL.

Calved 1881; breeder Mr. P. Blofield; s. Rollick 558; d. Rosette 1162 by Young Major 235; gr d. Rose by a Stoke Bull; 2nd gr d. Fuller—K 26 by a Stoke Bull.

734   REBEL.

Calved March 17, 1882; breeder Mr. A. Taylor; s. Starston Duke 570; d. Needful 1041 by Easton Duke 61; gr d. Nancy 363 by Richard II. 173; 2nd gr d. Lovely 2nd 322 by Richard II. 173; 3rd gr d. Pretty 425 by Richard I. 172; 4th gr d. Lily 313 by Laxfield Sire 101; 5th gr d. Lovely—R 2 by Laxfield Sire 101.

735   RED KNIGHT.

Calved July 31, 1880; breeder the Right Hon. Lord Henniker; s. Crown Prince 281; d. Strawberry—O 13.

736 RINALDO.

Calved July, 1881; breeder Mr. Fulcher; s. Tommy 588; d. Rose 1793 by Redhead 552; gr d. Cherry 94 by Fransham Captain 71; 2nd gr d. Tit—N 6 by Necton 3rd 122.

737 RINALDO.

Calved February 14, 1882; breeder Mr. G. F. Taber; owner Mr. D. L. Stevens; s. Champion 271; d. Sadie 1835 by Champion 271; gr d. Susie 1220—AN 7 by Ravinewood Beau 160; 2nd gr d. Skelton—N 7 by Necton 3rd 122.

738 ROGER.

Calved March 18, 1881; breeder Mr. G. F. Taber; owner Mr. H. J. Chamberlin; s. Ravinewood Beau 160; d. Cauliflower 3rd 82 by Shylock 166; gr d. Cauliflower 81 by Sampson 191; 2nd gr d. Primula—U 5.

739 ROLAND.

Calved December 30, 1882; breeder Mr. N. Powell; s. Premier 543; d. Ruby 1825 by Norfolk John 2nd 527; gr d. Ringlet 1783 by Norfolk John 131; 2nd gr d. Rose 481 by Redjacket 7th 169; 3rd gr d. Polly—P 7.

740 ROMANO.

Calved December 1881; breeder Mr. Fulcher; owner Sir E. C. Kerrison, Bart.; s. Lofty 515; d. Rosy Cross 1822 by Duke 52; gr d. Rose of Eaton 504 by Cringleford Sire 44; 2nd gr d. Cowslip 125 by Stoke 208; 3rd gr d. Rose 470 by Son of Hapton 205; 4th gr d. Cherry—E 2.

741 ROMEO.

Calved December 4, 1881; breeder Mr. J. J. Colman, M.P.; owner Col. J. B. Mead and Mr. R. J. Kimball; s. Rufus 188; d. Rosa 1133 by Norfolk Duke 127; gr d. Rose 3rd 480 by Young Duke 234; 2nd gr d. Rose 2nd 479 by Tenant Farmer 213; 3rd gr d. Rose—P 3.

742 ROMEO.

Calved October 26, 1882; breeder Mr. H. Biddell; s. Monarch 4th 351; d. Cherry Pie 787 by Earl of Suffolk 297; gr d. Cherry Lux—B 2.

564 ROUNDHEAD.

[*Corrected Entry.*]

Calved February 16, 1881; breeder Mr. J. J. Colman, M.P.; s. Rufus 188; d. Fan 1506 by Roundhead 180; gr d. Fanny 220 by Hero 3rd 87; 2nd gr d. Madame Freeman—A 9.

743 ROYAL DUKE.

Calved March 29, 1881; breeder Mr. N. Powell; s. Norfolk John 2nd 527; d. Ringlet 1783 by Norfolk John 131; gr d. Rose 481 by Redjacket 7th 169; 2nd gr d. Polly—P 7.

744 RULER OF ELMHAM.

Calved March 6, 1882; breeder Mr. Fulcher; s. Lofty 515; d. Ruby 1823 by Rufus 188; gr d. Ruddy 1831 by The Palmer 138; 2nd gr d. Fenn—A 37 by Hero 3rd 86.

745 RUPERT.

Calved August 27, 1882; breeder Mr. N. Powell; s. Premier 543; d. Rose Leaf 1817 by Norfolk John 2nd 527; gr d. Rose 481 by Redjacket 7th 169; 2nd gr d. Polly—P 7.

746 RUPERT.

Calved October 22, 1882; breeder the Right Hon. Lord Hastings; s. Roscoe 559; d. Davy 19th 848 by Davyson 3rd 48; gr d. Davy 12th 174 by The Baron 9; 2nd gr d. Davy 5th 167 by Tenant Farmer 213; 3rd gr d. Davy—H 1.

747 SANDBOY.

Calved March 4, 1882; breeder Mr. H. Biddell; owner Mr. B. M. Haggard; s. Monarch 4th 351; d. Currant Wine 1424 by Iron Duke 125; gr d. Cherry Pie 787 by Earl of Suffolk 297; 2nd gr d. Cherry Lux—B 2.

748 SAXHAM PRINCE.

Calved December, 1880; breeder Mr. John Jillings; owner the Most Hon. the Marquis of Bristol; s. Crown Prince 281; s. Saxham Waxwork by Nelson 357; gr d. Waxwork 3rd 1264 by King Theodore 98; 2nd gr d. Waxwork—U 9.

749 SCRANTON ÆSTHETE.

Calved September 14, 1881; breeder Mr. R. E. Lofft; owner Mr. J. E. Carmalt; s. Doubtful 487; d. Poppet 2nd 1087 by Cherry Duke 32; gr d. Poppet—U 43 by Sampson 191.

750 SILVERSIDES.

Calved May, 1881; breeder Mr. G. Gooderham; owner Mr. W. A. T. Amherst, M.P.; s. Troston 3rd 591; d. Silverlocks 551 by The Baron 10; gr d. Silverbury 550 by Playford Sire 142; 2nd gr d. Bury—B 10.

751 SIR DAVID.

Calved August 9, 1882; breeder Mr. W. A. T. Amherst, M.P.; s. Davyson 3rd 48; d. Nancy 2nd 1691 by Young Major 235; gr. d. Nancy 1690 by Peck 534; 2nd gr d. Spot 3rd 1863 by Wilby Chapman 228; 3rd gr d. Spot 558 by Wonder 231; 4th gr d. Rose—K 19 by an Elmham Bull.

752 SIR GARNET.

Calved May 11, 1882; breeder Mr. H. Biddell; owner the Rev. Charles Terry; s. Monarch 4th 351; d. Christmas Bloom 793 by Iron Duke 125; gr d. Rosebloom 485 by Seneca 195; 2nd gr d. Rosebud—B 21.

753 SIR NICHOLAS.

Calved August 26, 1882; breeder Mr. W. A. T. Amherst, M.P.; s. Davyson 3rd 48; d. Dolly 1464 by Norfolk John 131; gr d. Daisy 1436 by Redjacket 7th 169; 2nd gr d. Cherry—P 9.

754 SIR SAMUEL.

Calved September 21, 1882; breeder Mr. W. A. T. Amherst, M.P.; s. Shylock 572; d. Goldenlocks 1545 by Iron Duke 125; gr d. Silverlocks 551 by The Baron 10; 2nd gr d. Silverbury 550 by Playford Sire 142; 3rd gr d. Bury—B 10.

755 SIR SIMEON.

Calved September 13, 1882; breeder Mr. W. A. T. Amherst, M.P.; s. Shylock 572; d. Marjoram 1661 by Iron Duke 125; gr d. Wild Thyme 653 by The Baron 10; 2nd gr d. Wild Cherry 652; 3rd gr d. Cherry—V 5.

REGISTER OF BULLS.

756 SMALL.
Calved September 10, 1881; breeder Mr. R. E. Lofft; s. Bantam 451; d. Gloss 3rd 1542 by Bright 267; gr d. Gloss 2nd 665 by Boss 237; 2nd gr d. Gloss—V 11.

757 SMART.
Calved June 7, 1882; breeder Mr. R. E. Lofft; s. Long 516; d. Rosebud—I 13 by a Rudham Bull.

758 STARSTON PRINCE.
Calved December 2, 1882; breeder Mr. A. Taylor; s. Starston Duke 570; d. Nosegay 3rd 1055 by Harold 83; gr d. Nosegay 393 by Rifleman 175; 2nd gr d. Nosegay—O 5.

759 SUPERB.
Calved June 10, 1882; breeder Mr. G. F. Taber; owner Mr. J. T. Lewis; s. Powerful 728; d. Daisy 2nd 2121 by Hector 319; gr d. Daisy 156 by Sampson 191; 2nd gr d. Phœnix—U 6.

760 SWEETMEAT.
Calved September 16, 1882; breeder Mr. H. Biddell; s. Shylock 572; d. Honeysuckle 1574 by Iron Duke 125; gr d. June Rose 968 by The Baron 10; 2nd gr d. Rose—B 9.

761 SWIPES.
Calved January 8, 1883; breeder Mr. G. Holmes; s. Brundish Prince 462; d. Beauty 2005 by Harold 83; gr d. Nancy 2394 by Read 385; 2nd gr d. Brundish—R 9 by a Laxfield Bull.

762 TANCRED.
Calved February 24, 1882; breeder Mr. R. H. Mason; s. Philip 538; d. Tit 4th 1899 by Lord Easton 105; gr d. Tit 2nd 1897 by Lord Easton 105; 2nd gr d. Tit—N 6 by Necton 3rd 122.

763 THORNHAM HERO.
Calved May 11, 1880; breeder the Right Hon. Lord Henniker; s. Crown Prince 281; d. Daisy 2nd 828 by Eclipse 2nd 299; gr d. Daisy 827 by Eclipse 63; 2nd gr d. Strawberry—O 13.

REGISTER OF BULLS.

764 TIMON.
Calved October 13, 1882; breeder Mr. H. Biddell; s. Shylock 572; d. Trimley Cherry 1240 by Trimley 423; gr d. Cherry—V 5.

765 TIMON.
Calved December 16, 1882; breeder Mr. T. Leonard Palmer; s. Alonso 447; d. Sweet Pea 2567 by Bridegroom 2nd 630; gr d. Sweet Pea 595 by Waxwork 597; 2nd gr d. Sweet Pea—U 14.

766 TOM.
Calved February, 1882; breeder Mr. W. Bradfield; s. Tommy 588; d. Nettie 1046 by The Palmer 138; gr d. Norton 392 by Hero 3rd 87; 2nd gr d. Norton—A 6.

767 TRIMLEY JOHN.
Calved July 17, 1881; breeder Mr. C. K. Cordy; s. Ready 551; d. Trimley Daisy 1907 by Roarer 392; gr d. Daisy—X 4 by Zephyr 441; 2nd gr d. by Youngster 439.

768 TRIMLEY TRUE HEART.
Calved November 17, 1881; breeder Mr. H. Biddell; owner the Right Hon. the Earl of Stradbroke; s. St. Edmund 580; d. Trimley Cherry 1240 by Trimley 423; gr d. Cherry—V 5.

769 TROSTON 6TH.
Calved October 1881; breeder Mr. G. Gooderham; s. Wild Robin 600; d. Floss 2nd 1523 by Troston 424; gr d. Floss 900 by Perfection 140; 2nd gr d. Favourite 222 by Doncaster 50; 3rd gr d. Flora 229 by King Alfred 96; 4th gr d. Red Stockings—V 2 by Wonder 230.

770 TROSTON 7TH.
Calved September 3, 1882; breeder Mr. G. Gooderham; s. Wild Roger 603; d. Floss 2nd 1523 by Troston 424; gr d. Floss 900 by Perfection 140; 2nd gr d. Favourite 222 by Doncaster 50; 3rd gr d. Flora 229 by King Alfred 96; 4th gr d. Red Stockings—V 2 by Wonder 230.

771 TROSTON PRINCE.
Calved September 5, 1881; breeder Mr. Lofft; owner Mr. W. B. Easter; s. Doubtful 487; d. Bridesmaid 7th 1337 by Duke 488; gr d. Bridesmaid 2nd 721 by Rudham Hero 183; 2nd gr d. Bridesmaid—I 9.

772 UNCAS.

Calved June 30, 1882; breeder Mr. John Margarson; s. Purl 611; d. Una 2nd 1920 by Lord of the Manor 338; gr d. Una 1243 by The Freeman 309; 2nd gr d. Upton—L 3 by The Palmer 138.

773 VICTOR.

Calved August 25, 1882; breeder Mr. R. H. Mason; s. Slasher 577; d. Dainty 1427 by Osman 521; gr d. Violet 4th 1252 by Lord Easton 105; 2nd gr d. Dainty 140 by Prince Charlie 151; 3rd gr d. Nancy 359 by Fransham Captain 71; 4th gr d. Tit—N 6 by Necton 3rd 122.

774 VALOUR.

Calved July, 1881; breeder Mr. W. Harvey; owner Mr. R. L. Harvey; s. Victor 596; d. Ringlet—U 30 by Timworth 420.

775 VILLEBOIS.

Calved June 6, 1882; breeder Mr. Fulcher; s. Roscoe 559; d. Vestal 2594 by Brutus 269; gr d. Violet 2nd 1925 by The Palmer 138; 2nd gr d. Violet—A 26 by Hero 2nd 86.

776 WAGGONER.

Calved January 6, 1883; breeder Capt. J. Borlase Tibbits; s. Plowboy 540; d. Young Darkie 3rd 1950 by Ruler 463; gr d. Darkie 3rd 842 by Young Foxhall 437; 2nd gr d. Darkie 841 by a Suffolk Bull; 3rd gr d. Rose by a Suffolk Bull; 4th gr d. Gillyflower by a Suffolk Bull; 5th gr d. Buttercup—Y 1 by a Suffolk Bull.

777 WATCHMAN.

Calved 1877; breeder Mr. H. Biddell; owner Mr. D. Burrows; s. Crown Prince 281; d. Wise Woman 656 by Seneca 195; gr d. Fairy—B 17.

778 WILD RUFUS.

Calved April 29, 1881; breeder Mr. G. Gooderham; owner the Right Hon. Lord Suffield, K.C.B.; s. Troston 3rd 591; d. Wild Rose of Kilburn 1939 by Troston 424; gr d. Wild Rose 1271 by The Claimant 34; 2nd gr d. Rosy 513 by Perfection 140; 3rd gr d. Beauty 36 by Wonder 230; 4th gr d. Cowslip—V 1.

779 WILD RULER.

Calved December 18, 1881; breeder Mr. G. Gooderham; s. Wild Rocket 601; d. Wild Rose of Kilburn 1939 by Troston 424; gr d. Wild Rose 1271 by The Claimant 34; 2nd gr d. Rosy 513 by Perfection 140; 3rd gr d. Beauty 36 by Wonder 230; 4th gr d. Cowslip—V 1.

780 WISEACRE.

Calved October 16, 1880; breeder Mr. H. Biddell; owner Mr. F. D. Kent; s. Ironsides 509; d. Wisdom 656 by Seneca 195; gr d. Fairy—B 17.

781 YOUNG DUKE.

Calved April 8, 1881; breeder the Right Hon. Lord Hastings; s. Thornham Duke 2nd 585; d. Davy Duchess 1460 by Davyson 4th 286; gr d. Davy 16th 845 by Redjacket 7th 169; 2nd gr d. Davy 7th 169 by Young Duke 234; 3rd gr d. Davy 2nd 164 by Sir Nicholas 202; 4th gr d. Davy—H 1.

782 YOUNG RIVAL.

Calved September 25, 1881; breeder Mr. R. E. Lofft; s. Stout 581; d. Elmham Rosebud 2nd 872 by Prince Regent 381; gr d. Elmham Rosebud 195 by Hero 2nd 86; 2nd gr d. Rose by Redjacket 2nd 164; 3rd gr d. Primrose—A 1 by Elmham Sire 67.

783 Z.

Calved November 29, 1882; breeder Mr. H. Biddell; s. Shylock 572; d. China Rose 788 by The Baron 10; gr d. Rose—B 9.

784 ZERO.

Calved December 24, 1882; breeder Mr. H. Biddell; s. Shylock 572; d. Little Lady 1004 by The Baron 10; gr d. Lady—W 9.

---

740 ROMANO.
[*Corrected Entry.*]

Calved December 1881; breeder Mr. Fulcher; owner Sir E. C. Kerrison, Bart.; s. Lofty 515; d. Rosy Cross 1822 by Brutus 269; gr d. Rosette 1150 by Baker 253; 2nd gr d. Roseleaf 499 by Powell 143; 3rd gr d. Rose—E 5 by Cringleford Sire 44.

758   STARSTON PRINCE.
[*Corrected Entry.*]

Calved December 2, 1882; breeder Mr. A. Taylor; s. Starston Duke 570; d. Nosegay 2nd 394 by Major 109; gr d. Nosegay 393 by Rifleman 175; 2nd gr d. Nosegay—O 5.

CORRECTION:—In Pedigree of

    HAMLET   500

    IRON HEART 508

    STOUT HEART 582

For WHITEHEART by CREMORNE 42

Read WHITEHEART by SENECA 195.

In Pedigree of
ALDEBY 614 *read* NANCY 2394.

REGISTER OF COWS.

1967   ABBESS — E 2.

Calved January 28, 1881; breeder Mr. H. Waring; s. Rufus 188; d. Bessie 2nd 1321 by Lord Russell 342; gr d. Bessie 45 by Stoke Duke 209. See Little Bess—E 2 by Spot 206.

1968   ABBESS — M 1.

Calved December 24, 1880; breeder Mr. T. Brown; s. Priam 373; d. Agar 670 by Powell 143; gr d. Alma 15th 12 by Tenant Farmer 213. See Alma 12th—M 1 by Byron 28.

1969   ABIGAIL — E 12.

Calved March 16, 1881; breeder Mr. H. Waring; s. Robin Hood 394; d. Europa 1500 by Suffolk 211; gr d. Susanna 587 by Stoke Duke 209. See Susan—E 12.

1970   ACACIA — F 9.

Calved November 9, 1881; breeder Mr. H. Waring; s. Beau 259; d. Brenda 1332 by Disraeli 289; gr d. Libby 1619 by Cherry Duke 32. See Boo—F 9 by Baron Easton 11.

1971   ACACIA — M 1.

Calved January 29, 1882; breeder Mr. T. Brown; s. Priam 373; d. Agatha 1303 by Royal Duke 181; gr d. Alice 4 by The Peer 139. See Alma 15th—M 1 by Tenant Farmer 213.

1972 ACCEPTANCE — F 9.

Calved December 9, 1881; breeder Mr. H. Waring; s. Beau 259; d. Minna 1671 by Disraeli 289; gr d. Libby 1619 by Cherry Duke 32. See Boo—F 9 by Baron Easton 11.

1973 ACCIDENT — W 4.

Calved Febrnry 7, 1882; breeder Mr. H. Waring; s. Charles 469; d. Sphinx 1860 by Disraeli 289; gr d. Miss Colman 1024 by Mr. Boodybidums 118. See Violet 2nd—W 4 by Duke of Suffolk 57.

1974 ACCURATE — P 4.

Calved March 14, 1882; breeder Mr. H. Waring; s. Charles 469; d. Ethel 1498 by Roundhead 180; gr d. Blue Bell 52 by Norfolk Duke 127. See Nina—P 4 by Tenant Farmer 213.

1975 ACHIEVEMENT — F 3.

Calved May, 1882; breeder Mr. H. Waring; s. Haman 499; d. Joanna 283 by Cherry Duke 32; gr d. Joan—F 3.

1976 ACONITE — B 11.

Calved December 30, 1882; breeder the Right Hon. W. H. Smith, M.P.; s. Bluebeard 625; d. Belliza 1319 by Monarch 4th 351; gr d. Bellona 705 by The Baron 10. See Suffolk Belle—B 11 by Seneca 195.

1977 ACTRESS — M 1.

Calved February 12, 1881; breeder Mr. T. Brown; s. Priam 373; d. Alicia 1304 by Norfolk Duke 127; gr d. Alice 4 by The Peer 139. See Alma 15th—M 1 by Tenant Farmer 213.

1978 ADELAIDE — M 1.

Calved December 4, 1882; breeder Mr. T. Brown; s. Priam 373; d. Alice 4 by The Peer 139; gr d. Alma 15th 12 by Tenant Farmer 213. See Alma 12th—M 1 by Byron 28.

1979 AGNES — V 15.

Calved 1882; breeder Sir C. Rowley; owner Lieut.-Col. W. B. Long; s. a Tendring Bull; d. a Tendring Cow—V 15.

1980   ALFORATA — W 14.

Calved March 20, 1882; breeder Her Grace the Duchess of Hamilton; owner Messrs. J. H. and W. W. Clark; s. The Wilby Lad 599; d. Emmeline 2165 by Handsome Prince 317; gr d. Esmeralda 873 by Roundhead 180. See Emerald—W 14 by Stoke Duke 209.

1981   ALMA — P 4.

Calved February 2, 1881; breeder Mr. H. Waring; s. Rufus 188; d. Ethel 1498 by Roundhead 180; gr d. Blue Bell 52 by Norfolk Duke 127. See Nina—P 4 by Tenant Farmer 213.

1982   AMPTON — U 16.

Calved 1879; breeder Mr. Hunter Rodwell; owner Mr. H. Spurling; s. Bridegroom 2nd 630; d. Violet by Waxwork 597; gr d. Violet—U 16.

1983   ANNA — A 22.

Calved January 16, 1880; breeder the Right Hon. Lord Suffield, K.C.B.; s. Norfolk 361; d. Amy 674 by Rufus 187; gr d. Alice 671 by Witton 432. See Alice—A 22 by Hero of Newcastle 85.

1984   ANNIE — A 13.

Calved March 23, 1882; breeder Sir J. W. C. Hartopp, Bart.; s. Hardwick 501; d. Daisy 2118 by Bounty 460; gr d. Red Stocking 1778 by Rufus 188. See Spot—A 13.

1985   ANNIE — E 11.

Calved 1880; breeder Sir J. W. C. Hartopp, Bart.; s. Bounty 460; d. Gladiolus 1537 by Brutus 269; gr d. Pansy 1063 by Cringleford Duke 243. See Pretty—E 11 by Cantley 29.

1986   ANNIE — K 17.

Calved 1879; breeder Sir J. W. C. Hartopp, Bart.; s. Bounty 460; d. Bright Cherry 724 by Peter Powell 370; gr d. Cherry 4th 768 by Norfolk Duke 127. See Cherry 2nd—K 17 by Norfolk Duke 127.

1987   APPLE GIRL.

Probationer—Calved April 8, 1882; breeder Mr. Garrett Taylor; s. Davyson 8th 477; d. Appleton Brownie 2nd 1955 by an Elmham Bull.

REGISTER OF COWS.

1988 APPLE SAUCE.

Probationer—Calved May 17, 1881; breeder Mr. Garrett Taylor; s. Albert Victor (an Elmham Bull); d. Appleton Brownie 2nd 1955 by an Elmham Bull.

1989 ASHLEAF — A 33.

Calved March 11, 1881; breeder Mrs. E. Perkins; s. Osman 531; d. Elm-leaf 2nd 1489 by King Tom 335; gr d. Elm-leaf—A 33 by a Ramsley Bull.

1990 ASHLEAF 2ND — A 33.

Calved January 24, 1882; breeder Mrs. E. Perkins; s. Othello 532; d. Elm-leaf 2nd 1489 by King Tom 335; gr d. Elm-leaf—A 33 by a Ramsley Bull.

1991 AUBURN — M 1.

Calved November 19, 1881; breeder Mr. T. Brown; s. Priam 373; d. Alice 4 by The Peer 139; gr d. Alma 15th 12 by Tenant Farmer 213. See Alma 12th—M 1 by Byron 28.

1992 BAKER — 1 SUFF.

Calved 1873; breeder Mr. Baker; owner Mr. Geo. Gooderham; s. the late Mr. C. Austin's Blood Red Bull; d. by the Rev. G. T. Turner's Blood Red Bull; gr d. by Mr. C. Austin's Blood Red Bull. See Introduction.

1993 BALLAD — F 8.

Calved January 20, 1883; breeder Mr. H. Waring; s. Charles 469; d. Fuchsia 1528 by Suffolk 211; gr d. Fanny—F 8.

1994 BALY 1ST.

Probationer—Calved March 1881; breeder Mr. Jno. Baly; owner Mr. Fulcher; s. Brutus Duo 463; d. Cherry by an Elmham Bull.

1995 BALY 2ND.

Probationer—Calved April, 1881; breeder Mr. Jno. Baly; owner Mr. Fulcher; s. Brutus Duo 463; d. Dereham by an Elmham Bull.

1996   BARMAID – R 1.

Calved May 11, 1881; breeder Mr. H. Biddell; s. Monarch 4th 351; d. Susanna 1219 by Norfolk Duke 127; gr d. Susan 586 by Tommy 216. See Sarah—R 1 by Elmham 65.

1997   BARONESS DAVY – H 1.

Calved January 9, 1879; breeder the Right Hon. Lord Hastings; s. Davyson 4th 286; d. Davy 19th 848 by Davyson 3rd 48; gr d. Davy 12th 174 by The Baron 9. See Davy 5th—H 1 by Tenant Farmer 213.

1998   BATTERSEA 5TH – W 1.

Calved November, 1878; breeder Mr. H. Wolton; s. Perfection 2nd 368; d. Battersea 4th 689 by Oakley 133; gr d. Battersea 2nd 22 by Garibaldi 73. See Heiress—W 1 by Battersea Eclipse 14.

1999   BATTERSEA BELLE – W 1.

Calved November, 1880; breeder Mr. H. Wolton; s. May Prince 349; d. Battersea 4th 689 by Oakley 133; gr d. Battersea 2nd 22 by Garibaldi 73. See Heiress—W 1 by Battersea Eclipse 14.

2000   BAUBEE – W 4.

Calved January 19, 1883; breeder Mr. H. Waring; s. Charles 469; d. Sphinx 1860 by Disraeli 289; gr d. Miss Colman 1024 by Mr. Boodybidums 118. See Violet 2nd—W 4 by Duke of Suffolk 57.

2001   BEATRICE – A 24.

Calved June 17, 1881; breeder the Right Hon. Lord Suffield, K.C.B.; owner Mrs. Collyer; s. Rupert 567; d. Flower 901 by Rufus 187; gr d. Bridget 723 by Witton 432. See Floss—A 24 by Witton 432.

2002   BEAUTIFUL BELLE – W 2.

Calved June 15, 1882; breeder Mr. S. Wolton; s. Wild Rover 605; d. 8th Belle of Suffolk 1317 by Perfection 2nd 368; gr d. 5th Belle of Suffolk 44 by Duke of Suffolk 57. See 4th Belle of Suffolk—W 2 by Garibaldi 73.

2003   BEAUTY – B 11.

Calved August 15, 1882; breeder Mr. H. Biddell; owner Mr. W. A. T. Amherst, M.P.; s. Bluebeard 625; d. Beatrice 1308 by Monarch 4th 351; gr d. Bellona 705 by The Baron 10. See Suffolk Belle—B 11 by Seneca 195.

REGISTER OF COWS.

2004 BEAUTY — N 4.
Calved December 13, 1881; breeder Mrs. E. Perkins; s. Othello 532; d. Daisy 3rd 1434 by Bradfield 264; gr d. Daisy 152 by Necton 3rd 122. See Rose—N 4 by Necton Prize 120.

2005 BEAUTY — R 9.
Calved November 1877; breeder Mr. T. Easter; owner Mr. G. Holmes; s. Harold 83; d. Nancy 2394 by Read 385; gr d. Brundish—R 9 by a Laxfield Bull.

2006 BEAUTY — S 2.
Calved April 1, 1881; breeder Sir J. W. C. Hartopp, Bart.; s. Hardwick 501; d. Red Berry 1765 by Roundhead 180; gr d. Harebell 254 by Powell 143. See Holly—S 2 by Tommy 216.

2007 BEAUTY — 2 SUFF.
Calved July, 1880; breeder Mr. E. Boon; s. Troston 2nd 590; d. Cherry 2nd 2081 by a Thornham Bull; gr d. Cossett—2 SUFF. by an Oakley Bull. See Introduction.

2008 BECKY SHARPE — A 27.
Calved March 26, 1882; breeder Mr. Fulcher; s. Lofty 515; d. Brindy 729 by Hero 2nd 86; gr d. Curson—A 27 by Money 352.

2009 BELLA — B 11.
Calved February 2, 1882; breeder Mr. H. Biddell; s. Shylock 572; d. Bellona 705 by The Baron 10; gr d. Suffolk Belle 582 by Seneca 195. See Suffolk—B 11.

2010 BELLE — B 11.
Calved June 12, 1878; breeder Mr. H. Biddell; owner Mr. A. J. Smith; s. Iron Duke 125; d. Bellona 705 by The Baron 10; gr d. Suffolk Belle 582 by Seneca 195. See Suffolk—B 11.

2011 BELLE OF CHATHAM — AA 12.
Calved April 7, 1882; breeder Mr. G. F. Taber; owner Mr. J. T. Lewis; s. Champion 271; d. Mabel 1649 by General 496; gr d. May 1015—AA 12 by Ravinewood Beau 160. See Ocean Maid—A 12 by Hero 3rd 87.

REGISTER OF COWS.

2012 BELLE OF THE SANDS — W 2.
Calved April 3, 1882; breeder Mr. S. Wolton; s. Wild Rover 605; d. 7th Belle of Suffolk 1316 by Oakley 133; gr d. 5th Belle of Suffolk 44 by Duke of Suffolk 57. See 4th Belle of Suffolk—W 2 by Garibaldi 73.

2013 10TH BELLE OF SUFFOLK — W 2.
Calved July 9, 1881; breeder Mr. S. Wolton; s. Perfection 2nd 368; d. 6th Belle of Suffolk 704 by Oakley 133; gr d. 5th Belle of Suffolk 44 by Duke of Suffolk 57. See 4th Belle of Suffolk—W 2 by Garibaldi 73.

2014 11TH BELLE OF SUFFOLK — W 2.
Calved December 12, 1881; breeder Mr. S. Wolton; s. Perfection 2nd 368; d. 5th Belle of Suffolk 44 by Duke of Suffolk 57; gr d. 4th Belle of Suffolk 43 by Garibaldi 73. See Topsy—W 2 by Robinson 178.

2015 12TH BELLE OF SUFFOLK — W 2.
Calved July 30, 1882; breeder Mr. S. Wolton; s. Wild Rover 605; d. 6th Belle of Suffolk 704 by Oakley 133; gr d. 5th Belle of Suffolk 44 by Duke of Suffolk 57. See 4th Belle of Suffolk—W 2 by Garibaldi 73.

2016 BERENICE — A 27.
Calved November 5, 1882; breeder Mr. W. A. T. Amherst, M.P.; s. Davyson 3rd 48; d. Bertha 1320 by Brutus 269; gr d. Brindy 2nd 1341 by Rufus 188. See Brindy—A 27 by Hero 2nd 86.

2017 BERENICE — N 2.
Calved February 9, 1881; breeder Mr. T. Brown; s. Priam 373; d. The Elmham Belle 202 by Hero 2nd 86; gr d. Minnie—N 2 by Necton Prize 120.

2018 BETSY — W 3.
Calved December 13, 1882; breeder Mr. S. Wolton; s. Wild Rover 605; d. Nelly 4th 1045 by Oakley 133; gr d. Nelly 3rd 377 by Rendlesham Hero 171. See Nelly 2nd—W 3 by Prince Regent 153.

2019 BEZIQUE — B 20.
Calved August 25, 1881; breeder Mr. H. Biddell; s. Monarch 4th 351; d. Piquet 1076 by The Baron 10; gr d. Picket—B 20.

REGISTER OF COWS.

2020   BLACK BLOSSOM — B 13.

Calved May 1, 1878; breeder Mr. H. Biddell; owner Mr. A. J. Smith; s. Iron Duke 125; d. Blossom—B 13 by Powell's of Kelvedon.

2021   BLACKING — T 6.

Calved January 4, 1882; breeder Mr. H. Birkbeck; s. Haman 499; d. B.B. 1315 by Beau 259; gr d. Bee 77 by Young Duke 234. See Brownie—T 6 by Tenant Farmer 213.

2022   BLANCH — N 2.

Calved January 2, 1881; breeder Mr. T. Brown; s. Priam 373; d. Bianca 710 by Norfolk Duke 127; gr d. The Elmham Belle 202 by Hero 2nd 86. See Minnie—N 2 by Necton Prize 120.

2023   BLOSSOM — A 1.

Calved October, 1876; breeder Mr. W. Bradfield; owner Mr. H. Haylock; s. Rufus 188; d. Nellie 1702 by The Palmer 138; gr d. Nelly 871 by Hero 2nd 86. See Primrose—A 1 by Elmham Sire 67.

2024   BLOSSOM — A 5.

Calved 1873; breeder Mr. W. Bradfield; owner Rev. H. Evans Lombe; s. a Ramsley Bull; d. Ramsley—A 5 by Hero of Newcastle 85.

2025   BLOSSOM — B 21.

Calved August 21, 1881; breeder the Most Hon. the Marquis of Bristol; s. Fancy King 491; d. Full Bloom 1529 by Iron Duke 125; gr d. Rose Bloom 485 by Seneca 195. See Rosebud—B 21.

2026   BLOSSOM 2ND — N 1.

Calved April 17, 1881; breeder Mrs. E. Perkins; s. Osman 531; d. Blossom 1323 by King Cole 330; gr d. Blossom by Prince Charlie 151. See Violet—N 1 by Necton 3rd 122.

2027   BLOSSOM — X 3.

Calved May 10, 1882; breeder Her Grace the Duchess of Hamilton; s. The Suffolk Baronet 583; d. Camelia 742 by Prince Arthur 150; gr d. Lovely by Plowman 371. See Cossett—X 3.

REGISTER OF COWS.

2028   BLUE BELL – P 4.

Calved September, 1877; breeder Mr. H. Birkbeck; s. Roundhead 180; d. Blue Bell 52 by Norfolk Duke 127; gr d. Nina 2nd 389 by Tenant Farmer 213. See Nina—P 4.

2029   BLUEBERRIES – A 27.

Calved May, 1880; breeder Mr. T. Fulcher; owner Mr. W. A. T. Amherst, M.P.; s. Brutus 269; d. Brindy 2nd 1341 by Rufus 188; gr d. Brindy 729 by Hero 2nd 86. See Curson—A 27 by Money 352.

2030   BLUEBERRIES 2ND – A 27.

Calved January 16, 1883; breeder Mr. W. A. T. Amherst, M.P.; s. Davyson 3rd 48; d. Blueberries 2029 by Brutus 269; gr d. Brindy 2nd 1341 by Rufus 188. See Brindy—A 27 by Hero 2nd 86.

2031   BLUE BONNET – P 4.

Calved November 1, 1881; breeder Mr. H. Birkbeck; s. Haman 499; d. Blue Bell 2028 by Roundhead 180; gr d. Blue Bell 52 by Norfolk Duke 127. See Nina 2nd—P 4 by Tenant Farmer 213.

2032   BLUSH ROSE – P 3.

Calved December 1, 1881; breeder Mr. Garrett Taylor; s. Grey Spot 498; d. Rose 5th 1146 by Norfolk Duke 127; gr d. Rose 2nd 479 by Tenant Farmer 213. See Rose—P 3.

2033   BLYTH.

Probationer—Calved November, 1882; breeder Mr. Blyth (Butley); owner Mr. Fulcher; s. Lofty 515; d. Blyth's Cow.

2034   BOO 4TH – F 9.

Calved December, 1882; breeder Mr. C. S. Read; s. Haman 499; d. Boo 2nd 1330 by Disraeli 289; gr d. Bones 715 by Baron Easton 11. See Boo—F 9.

2035   BOUNCE – A 12.

Calved June 27, 1882; breeder Mr. Jno. Howling; s. The Parson 533; d. Boulter 54 by Hero 3rd 87; gr d. Handsome—A 12.

2036   BOUNTIFUL – F 2.

Calved March 25, 1882; breeder Mr. B. Stimpson; s. Robin Hood 394; d. Beauty 692 by Rufus 189; gr d. Buttercup—F 2.

2037   BRACELET – W 14.

Calved January 20, 1882; breeder Her Grace the Duchess of Hamilton; s. The Suffolk Baronet 583; d. Esmeralda 873 by Roundhead 180; gr d. Emerald 204 by Stoke Duke 209. See Clara—W 14.

2038   BRIDESMAID – E 11.

Calved October 25, 1881; breeder Mr. W. B. Easter; owner Mr. F. Morris; s. Brundish Prince 462; d. Nelly 1704 by Harold 83; gr d. Polly—E 11 by Duke 52.

2039   BRIDESMAID 8TH – I 9.

Calved November 26, 1880; breeder Mr. R. E. Lofft; s. Prodigal 546; d. Bridesmaid 6th 1336 by Donald 291; gr d. Bridesmaid 3rd 722 by Cherry Duke 32. See Bridesmaid—I 9.

2040   BRIDESMAID 3RD – W 2.

Calved May 25, 1882; breeder Mr. S. Wolton; s. Wild Rover 605; d. Bridecake 720 by Oakley 133; gr d. Bridesmaid 59 by Duke of Suffolk 58. See 3rd Belle of Suffolk—W 2 by Nonpareil 126.

2041   BRIDGET OF TROSTON – I 9.

Calved September 11, 1881; breeder Mr. Lofft; owner Mr. Fulcher; s. Stout 581; d. Bridesmaid 2nd 721 by Rudham Hero 183; gr d. Bridesmaid—I 9.

2042   BRIGHT SPARK – A 3.

Calved April 4, 1881; breeder Mr. J. J. Colman, M.P.; owner Mr. Garrett Taylor; s. Rufus 188; d. Bright Lass 725 by Prince 148; gr d. Bright Lady 346 by The Palmer 138. See Lady Sondes—A 3 by Norfolk Duke 127.

2043   BROWN LOO – B 20.

Calved July, 1882; breeder Mr. A. J. Smith; s. Pickwick 720; d. Loo 2324 by Iron Duke 125; gr d. Piquet 1076 by The Baron 10. See Picket—B 20.

2044 BRUNETTE — P 3.

Calved June 24, 1882; breeder Mr. J. J. Colman, M.P.; s. King Charles 329; d. Brown 1343 by Duke of Norfolk 295; gr d. Isabelle 2nd 956 by Norfolk Duke 127. See Isabella—P 2 by Young Duke 234.

2045 BRUSH — P 2.

Calved November 4, 1881; breeder Mr. H. Birkbeck; s. Haman 499; d. Broom 731 by Trimmer 218; gr d. Bonnie 53 by Norfolk Duke 127. See Rose 2nd—P 3 by Tenant Farmer 213.

2046 BUFFET — K 19.

Calved April 5, 1882; breeder Mr. A. Taylor; s. Starston Duke 570; d. Buxom 1355 by Davyson 3rd 48; gr d. Cheerful 762 by Young Major 235. See Spot—K 19 by Wonder 231.

2047 BUSHY BLOSSOM — B 13.

Calved 1881; breeder Mr. A. J. Smith; s. Pickwick 720; d. Black Blossom 2020 by Iron Duke 125; gr d. Blossom—B 13 by Powell's of Kelvedon.

2048 BUSY BEE — B 12.

Calved December 16, 1879; breeder Her Grace the Duchess of Hamilton; owner Mr. Fulcher; s. Handsome Prince 317; d. Little Bee 1003 by Crown Prince 281; gr d. Queen Bee 480 by Seneca 195. See The Bee—B 12.

2049 BUTLEY BELLE — W 2.

Calved March 21, 1882; breeder Mr. S. Wolton; s. Perfection 2nd 368; d. 4th Belle of Suffolk 43 by Garibaldi 73; gr d. Topsy 611 by Robinson 178. See Belle of Suffolk—W 2 by Orwell 135.

2050 BUTTERCUP — I 13.

Calved December 4, 1881; breeder Mr. R. E. Lofft; owner the Right Hon. W. H. Smith, M.P.; s. Ross 562; d. Rosebud 4th 1799 by Prince 377; gr d. Rosebud 2nd 1153 by Rudham Hero 183. See Rosebud—I 13.

†346 BUTTERCUP — P 4.

[*Corrected Entry.*]

Calved November 22, 1880; breeder Mr. H. Birkbeck; s. Osman 530; d. Blue Bell 2028 by Roundhead 180; gr d. Blue Bell 52 by Norfolk Duke 127. See Nina 2nd—P 4 by Tenant Farmer 213.

2051   BUTTERFLY 7TH — Y 3.

Calved April 9, 1882; breeder Capt. J. Borlase Tibbits; s. Ruler 403; d. Butterfly 4th 1351 by Young Foxhall 437; gr d. Butterfly by Prince Leopold 380. See Wall-Eye 2nd—Y 3.

2052   BYLAUGH TIT — N 6.

Calved February 17, 1881; breeder Rev. H. Evans Lombe; s. Lord Elmham 519; d. Tit 2nd 1897 by Lord Easton 105; gr d. Tit—N 6 by Necton 3rd 122.

2053   BYLAUGH TIT 2ND — N 6.

Calved January 31, 1882; breeder Rev. H. Evans Lombe; s. Lord Elmham 519; d. Tit 2nd 1897 by Lord Easton 105; gr d. Tit—N 6 by Necton 3rd 122.

2054   CACTUS — A 4.

Calved February 3, 1882; breeder Mr. T. Brown; s. Priam 373; d. Camilla 1358 by Norfolk Duke 127; gr d. Kate 284 by Tenant Farmer 213. See Ringlet—A 4 by Hero of Newcastle 85.

2055   CAMELIA 2ND — X 3.

Calved May 1, 1881; breeder Her Grace the Duchess of Hamilton; owner Mr. G. F. Taber; s. Handsome Prince 317; d. Camelia Bud 743 by a Monewden Bull; gr d. Camelia 742 by Prince Arthur 150. See Lovely—X 3 by Plowman 371.

2056   CARELESS — P 9.

Calved June 6, 1882; breeder Mr. N. Powell; s. Premier 523; d. Cherry 2nd 1377 by Norfolk John 131; gr d. Cherry—P 9.

2057   CASSANDRA — A 4.

Calved August 5, 1882; breeder Mr. J. L. Marriott; s. Priam 373; d. Kitten 1608 by Royal Duke 181; gr d. Kathleen 972 by The Peer 139. See Ringlet 2nd—A 4 by Tenant Farmer 213.

2058   CASSIE — L 9.

Calved June 1881; breeder Mr. Jno. Margarson; s. Purl 611; d. Cheerly 1370 by Master Freeman 347; gr d. Cheerful 763 by Master Freeman 347. See Cherry—L 9.

2059   CECILIA — AU 5.

Calved March 13, 1880; breeder Mr. G. F. Taber; s. Ravinewood Beau 160; d. Cauliflower 3rd 82 by Shylock 196; gr d. Cauliflower 81 by Sampson 191. See Primula—U 5.

2060   CECILIA 2ND — AU 5.

Calved April 30, 1882; breeder Mr. G. F. Taber; owner Mr. E. W. English; s. Pomp 541; d. Cecilia 2059—AU 5 by Ravinewood Beau 160; gr d. Cauliflower 3rd 82 by Shylock 196. See Cauliflower—U 5 by Sampson 191.

2061   CELERY — A 33.

Calved May, 1880; breeder Mr. R. H. Mason; owner Mr. G. K. Taber; s. Popgun 542; d. Cherry of Necton 1390 by King Cole 330; gr d. Elm-leaf 2nd 1489 by King Tom 335. See Elm-leaf—A 33 by a Ramsley Bull.

2062   CELIA — V 5.

Calved December 7, 1882; breeder Mr. H. Biddell; s. Shylock 572; d. Juliet 1603 by Monarch 4th 351; gr d. Trimley Cherry 1240 by Trimley 423. See Cherry—V 5.

2063   CELLARETTE — A 33.

Calved July, 1881; breeder Mr. R. H. Mason; s. Slasher 577; d. Cherry of Necton 1390 by King Cole 330; gr d. Elm-leaf 2nd 1489 by King Tom 335. See Elm-leaf—A 33 by a Ramsley Bull.

2064   CHALICE — L 9.

Calved April, 1882; breeder Mr. John Margarson; s. Purl 611; d. Cherry 3rd 1376 by Lord of the Manor 338; gr d. Cherry—L 9.

2065   CHANTRESS — A 4.

Calved July 31, 1882; breeder Mr. T. Brown; s. Priam 373; d. Camlet 1359 by Royal Duke 181; gr d. Ringlet 2nd 465 by Tenant Farmer 213. See Ringlet—A 4 by Hero of Newcastle 85.

2066   CHARITY — A 4.

Calved January 29, 1881; breeder Mr. T. Brown; s. Priam 373; d. Ringlet 2nd 465 by Tenant Farmer 213; gr d. Ringlet 464 by Hero of Newcastle 85. See Brettenham Strawberry—A 4 by Redjacket 163.

REGISTER OF COWS.

2067 CHARITY – K 23.
Calved February 2, 1883; breeder Mr. T. Leonard Palmer; s. Rollick 558; d. Cherry 1374 by Young Major 235; gr d. Kate by Wonder 231. See Kate—K 23 by an Elmham Bull.

2068 CHARLOTTE – O 14.
Calved March 31, 1881; breeder the Right Hon. Lord Henniker; owner Mr. Fulcher; s. Cyprus 473; d. Cowslip by Eclipse 63; gr d. Cherry—O 14.

2069 CHARLOTTE – 1 NORF.
Calved October, 1881; breeder Mr. J. Rivett; s. Falstaff 303; d. Cherry 2080 by an Elmham Bll; gr d. a Pond Cow—1 NORF.

2070 CHARMER 4TH – K 18.
Calved October 11, 1881; breeder the Right Hon. Lord Hastings; s. Davyson 6th 475; d. Charmer 3rd 1368 by Davyson 3rd 48; gr d. Charmer 757 by Young Major 235. See Cherry—K 18 by Wonder 231.

2071 CHARMER – L 9.
Calved April, 1882; breeder Mr. John Margarson; s. Purl 611; d. Cheerly 1370 by Master Freeman 347; gr d. Cheerful 763 by Master Freeman 347. See Cherry—L 9.

2072 CHARMER – W 3.
Calved January 23, 1883; breeder Mr. B. Stimpson; s. Robin Hood 394; d. Cheerful 764 by Rufus 187; gr d. Cherry 2nd 101 by Cherry Duke 32. See Cherry—W 3 by Duke of Suffolk 56.

2073 CHARMER – 1 NORF.
Calved October, 1882; breeder Mr. W. Bradfield; s. Tommy 588; d. Wiffin Cherry 2606 by an Elmham Bull; gr d. a Pond Cow—1 NORF.

2074 CHARMING – K 19.
Calved June 16, 1881; breeder Mr. W. A. T. Amherst, M.P.; s. Davyson 3rd 48; d. Cheerful 762 by Young Major 235; gr d. Spot 558 by Wonder 231. See Rose—K 19 by an Elmham Bull.

2075   CHASTE — W 3.

Calved November 11, 1881; breeder Mr. B. Stimpson; s. Robin Hood 393; d. Cherry Pie 1385 by Robin Hood 394; gr d. Cherry 2nd 101 by Cherry Duke 32. See Cherry—W 3 by Duke of Suffolk 56.

2076   CHAZALIE — K 19.

Calved August 12, 1882; breeder Mr. W. A. T. Amherst, M.P.; s. Davyson 3rd 48; d. Cheerful 762 by Young Major 235; gr d. Spot 558 by Wonder 231. See Rose—K 19 by an Elmham Bull.

2077   CHEERING — K 19.

Calved June 16, 1881; breeder Mr. W. A. T. Amherst, M.P.; s. Davyson 3rd 48; d. Cheerful 762 by Young Major 235; gr d. Spot 558 by Wonder 231. See Rose—K 19 by an Elmham Bull.

2078   CHERRY 2ND — O 1.

Calved 1878; breeder Sir E. C. Kerrison, Bart.; s. Baronet 2nd 257; d. Cherry 771 by Harold 83; gr d. Victoria 625 by Rifleman 175. See Oakley—O 1 by Rifleman 175.

2079   CHERRY 2ND — R 10.

Calved December, 1876; breeder Mr. T. Easter; owner Mr. Burton; s. Harold 83; d. Cherry—R 10 by a Laxfield Bull.

2080   CHERRY — 1 NORF.

Calved 1871; breeder Mr. Wiffen; owner Mr. J. Rivett; s. an Elmham Bull; d. a Pond Cow—1 NORF. See Introduction.

2081   CHERRY 2ND — 2 SUFF.

Calved 1877; breeder Mr. E. Boon; s. a Thornham Bull; d. Cossett by an Oakley Bull; gr d. Nancy by a Kettleburgh (Turner) Bull. See Introduction.

2082   CHERRY OF ELMHAM — N 6.

Calved February, 1881; breeder Mr. Fulcher; s. Redhead 2nd 553; d. Cherry Bloom 1381 by Rufus 188; gr d. Cherry 94 by Fransham Captain 71. See Tit—N 6 by Necton 3rd 122.

2083   CHERRY BLOOM — B 2.

Calved November 8, 1880; breeder the Most Hon. the Marquis of Bristol; s. Rosario 560; d. Cherry Bud 781 by The Baron 10; gr d. Cherry Lux—B 2.

2084   CHERRY BLOSSOM — V 5.

Calved February 28, 1881; breeder the Most Hon. the Marquis of Bristol; s. Rosario 560; d. Cherry Duchess 1382 by Iron Duke 125; gr d. Double Cherry 855 by The Baron 10. See Cherry—V 5.

2085   CHERRY QUEEN — V 3.

Calved May 16, 1882; breeder Mr. W. A. T. Amherst, M.P.; s. Troston 3rd 591; d. Cherry Leaf 1384 by Troston 424; gr d. Cherry 98 by King Alfred 96. See Fillpail—V 3 by Wonder.

2086   CHERRY RIPE — K 17.

Calved November 6, 1882; breeder Mr. J. J. Colman, M.P.; s. King Charles 329; d. Cherry-leaf 1383 by The Beau 259; gr d. Cherry 5th 769 by Norfolk Duke 127. See Cherry 2nd—K 17 by Norfolk Duke 127.

2087   CHERRY ROSE — O 14.

Calved January 31, 1880; breeder the Right Hon. Lord Henniker; owner Mr. A. J. Smith; s. Crown Prince 281; d. Rose-leaf 1159 by Ruddy 402. See Cherry—O 14.

2088   CHLOE 6TH — W 21.

Calved May 13, 1881; breeder Mr. W. G. Collins; owner Mr. G. J. Paine; s. Hunston Duke 505; d. Chloe 4th by Prince 376; gr d. Chloe 2nd 790 by Cordy 274. See Chloe—W 21 by Rendlesham Hero 171.

2089   CHRISTINA 2ND — AU 5.

Calved February 6, 1882; breeder Mr. G. F. Taber; s. Champion 271; Christina 792—AU 5 by Rufus 188; gr d. Cauliflower 3rd 82 by Shylock 196. See Cauliflower—U 5 by Sampson 191.

2090   CICELY — A 4.

Calved August 21, 1882; breeder Mr. T. Brown; s. Priam 373; d. Katinska 973 by The Peer 139; gr d. Kate 284 by Tenant Farmer 213. See Ringlet—A 4 by Hero of Newcastle 85.

REGISTER OF COWS.

2091   CINNAMON – A 4.
Calved October 4, 1881; breeder Mr. T. Brown; s. Priam 373; d. Katinska 973 by The Peer 139; gr d. Kate 284 by Tenant Farmer 213. See Ringlet—A 4 by Hero of Newcastle 85.

2092   CLAREA – G 13.
Calved February 17, 1882; breeder Sir J. W. C. Hartopp, Bart.; s. Hardwick 501; d. Constance 2100 by Bounty 460; gr d. Castleacre 1364 by Fransham 445. See Golden Drop—G 13.

2093   CLARET 2ND – AF 6.
Calved May 12, 1882; breeder Mr. G. F. Taber; s. Robin Hood 394; d. Claret 1396 by Robin Hood 394; gr d. Clara—F 6.

2094   CLARKE 1ST.
Probationer—Calved August, 1880; breeder Mr. Brightmore Clarke; owner Mr. Fulcher; s. Brutus 269; d. Clarke's Cow.

2095   CLARKE 2ND.
Probationer—Calved September, 1881; breeder Mr. Brightmore Clarke; owner Mr. Fulcher; s. Lofty 515; d. Clarke's Cow.

2096   CLOISTER – L 9.
Calved May 17, 1882; breeder Mr. John Margarson; s. Slasher 577; d. Cherry 2nd 1375 by Master Freeman 347; gr d. Cherry—L 9.

2097   COLUMBINE – A 4.
Calved December 29, 1882; breeder Mr. T. Brown; s. Priam 373; d. Camilla 1358 by Norfolk Duke 127; gr d. Kate 284 by Tenant Farmer 213. See Ringlet—A 4 by Hero of Newcastle 85.

2098   COMFIT – L 9.
Calved March, 1882; breeder Mr. John Margarson; s. Purl 611; d. Cheerful 763 by Master Freeman 347; gr d. Cherry—L 9.

2099   CONGRESS – G 13.
Calved June 2, 1881; breeder Sir J. W. C. Hartopp, Bart.; s. Hardwick 501; d. Castleacre 1364 by Fransham 445; gr d. Golden Drop—G 13.

2100   CONSTANCE – G 13.

Calved 1879; breeder Sir J. W. C. Hartopp, Bart.; s. Bounty 460; d. Castleacre 1364 by Fransham 445; gr d. Golden Drop—G 13.

2101   COPFORD ROSE – O 3.

Calved May 14, 1881; breeder Mr. A. Taylor; owner Mr. T. H. Harrison; s. King Charles 329; d. Summer Rose 2559 by Harold 83; gr d. Cossett 1405 by Rifleman 175. See Cowslip—O 3 by Bowbearer 22.

2102   CORAL – B 17.

Calved November 4, 1882; breeder Mr. H. Biddell; s. Shylock 572; d. Cinderella 1394 by Crown Prince 281; gr d. Fairy—B 17.

2103   COSSETT 2ND – O 3.

Calved February 17, 1880; breeder Mr. A. Taylor; s. King Charles 329; d. Cossett 1405 by Rifleman 175; gr d. Cowslip—O 3 by Bowbearer 22.

2104   COSSETT 2ND – 2 SUFF.

Calved 1880; breeder Mr. E. Boon; s. Troston 2nd 590; d. Cossett by an Oakley Bull; gr d. Nancy by Kettleburgh Bull. See Introduction.

2105   COUNTESS – P 1.

Calved July, 1882; breeder Sir E. C. Kerrison, Bart.; s. Lofty 515; d. Duchess of Eye 2nd 863 by Prince 147; gr d. Duchess of Eye 190 by Farmer 70. See Duchess—P 1 by Tenant Farmer 213.

2106   COUNTESS – P 9.

Calved September 2, 1882; breeder Mr. N. Powell; s. Premier 543; d. Comassie 1957 by Norfolk John 131; gr d. Lady by Redjacket 7th 169. See Cherry—P 9.

2107   COUNTESS – U 45.

Calved January 13, 1882; breeder Mr. W. G. Collins; owner Mr. C. Austin; s. Stout 581; d. Constance 2nd 800 by Cherry Duke 32; gr d. Constance 799 by Plowman 371. See Weazel—U 45 by Newcastle Prize 359.

2108   COUSIN – O 3.

Calved April 8, 1881; breeder Mr. A. Taylor; s. King Charles 329; d. Cossett 1405 by Rifleman 175; gr d. Cowslip—O 3 by Bowbearer 22.

2109   COWSLIP — A 21.

Calved February 16, 1881; breeder the Right Hon. Lord Suffield, K.C.B.; owner Mr. C. K. Cordy; s. Rupert 567; d. Rosa 1131 by Rufus 187; gr d. Rose—A 21 by Hero of Newcastle 85.

2110   COWSLIP — P 9.

Calved April 8, 1881; breeder Mr. N. Powell; s. Norfolk John 2nd 527; d. Cherry 2nd 1377 by Norfolk John 131; gr d. Cherry—P 9.

2111   COWSLIP 3RD — T 15.

Calved January 5, 1882; breeder Mr. John Howell; s. Davyson 9th 478; d. Cowslip 2nd 1416 by Sir Thomas 409; gr d. Cowslip 811 by Rupert 405. See Buttercup—T 15 by Tenant Farmer 213.

2112   CRESS.

Probationer—Calved April 9, 1882; breeder Mr. Garrett Taylor; s. Davyson 8th 477; d. Cresswell Brownie 2nd 1955.

2113   CRESSIDA — A 4.

Calved August 7, 1881; breeder Mr. J. L. Marriott; s. Priam 373; d. Kitten 1608 by Royal Duke 181; gr d. Kathleen 972 by The Peer 139. See Ringlet 2nd—A 4 by Tenant Farmer 213.

2114   CROCUS — I 16.

Calved April, 1881; breeder Mr. W. Hudson; s. The Doctor 486; d. Countess 1406 by Davyson 5th 287; gr d. Countess by Quarles Duke 548. See Coral—I 16 by Proud 547.

2115   CROP-EARS — N 6.

Calved December, 1882; breeder Mr. Fulcher; s. The Parson 533; d. Cherry Bloom 1381 by Rufus 188; gr d. Cherry 94 by Fransham Captain 71. See Tit—N 6 by Necton 3rd 122.

2116   CURSON CHERRY — A 27.

Calved July 28, 1882; breeder Rev. A. G. Legge; s. The Parson 533; d. Curson—A 27 by Money 352.

2117   DAIRYMAID — 1 NORF.

Calved November 1, 1882; breeder Mr. J. Rivett; s. Falstaff 303; d. Dapple 2127 by an Elmham Bull; gr d. a Pond Cow.

2118   DAISY – A 13.

Calved 1879; breeder Sir J. W. C. Hartopp, Bart.; s. Bounty 460; d. Red Stocking 1778 by Rufus 188; gr d. Spot—A 13.

2119   DAISY – I 9.

Calved October 27, 1881; breeder Mr. R. E. Lofft; owner the Right Hon. W. H. Smith, M.P.; s. Ross 562; d. Bridesmaid 4th 1334 by Bernard 260; gr d. Bridesmaid—I 9.

2120   DAISY – O 7.

Calved June 5, 1882; breeder Her Grace the Duchess of Hamilton; s. The Wilby Lad 599; d. Daisy Girl 1439 by Roundhead 400; gr d. Cherry Cheeks 784 by Major 109. See Daisy—O 7.

2121   DAISY 2ND – U 6.

Calved September 28, 1881; breeder Mr. R. E. Lofft; s. Hector 319; d. Daisy 156 by Sampson 191; gr d. Phœnix—U 6.

2122   DAISY OF RAVINEWOOD – U 6.

Calved September 17, 1880; breeder Mr. R. E. Lofft; owner Mr. G. F. Taber; s. Hector 319; d. Daisy 156 by Sampson 191; gr d. Phœnix—U 6.

2123   DAISY CHAIN – H 2.

Calved July 30, 1881; breeder Mr. Garrett Taylor; s. Rufus 188; d. Daisy-leaf 1440 by Rufus 188; gr d. Daisy 3rd 823 by Powell 143. See Daisy 1st—H 2 by Young Duke 234.

2124   DAMSON – S 3.

Calved March 20, 1882; breeder Mr. Garrett Taylor; s. Cato 468; d. Damsel 1441 by Osman 530; gr d. Dainty 1428 by Powell 143. See Dorothy—S 3 by George of Elmham 76.

2125   DAMSON – 1 NORF.

Calved August, 1880; breeder Mr. J. Rivett; s. Falstaff 303; d. Dapple 2127 by an Elmham Bull; gr d. a Pond Cow.

2126   DANCER – 1 NORF.

Calved November, 1881; breeder Mr. J. Rivett; s. Falstaff 303; d. Dapple 2127 by an Elmham Bull; gr d. a Pond Cow.

2127   DAPPLE — 1 Norf.

Calved 1876; breeder Mr. Wiffen; owner Mr. J. Rivett; s. an Elmham Bull; d. a Pond Cow. See Introduction.

2128   DARK BEAUTY — Y 1.

Calved December 12, 1881; breeder Capt. J. Borlase Tibbits; s. Ruler 403; d. Young Darkie 2nd 1949 by Young Foxhall 437; gr d. Darkie 841 by a Suffolk Bull. See Rose—Y 1.

1457   DAVY 33RD — H 1.
[Corrected Entry.]

Calved 1880; breeder Mr. Jno. Hammond; s. Davyson 7th 476; d. Davy 21st 1445 by Davyson 5th 287; gr d. Daisy 7th 169 by Young Duke 234. See Davy 2nd—H 1 by Sir Nicholas 202.

2129   DAVY 36TH — H 1.

Calved June, 1881; breeder Mr. Jno. Hammond; s. Davyson 7th 476; d. Davy 7th 169 by Young Duke 234; gr d. Davy 2nd 164 by Sir Nicholas 202. See Davy—H 1.

2130   DAVY 37TH — H 1.

Calved June, 1881; breeder Mr. Jno. Hammond; s. Davyson 7th 476; d. Davy 21st 1445 by Davyson 5th 287; gr d. Davy 7th 169 by Young Duke 234. See Davy 2nd—H 1 by Sir Nicholas 202.

2131   DAVY 38TH — H 1.

Calved September 14, 1881; breeder Mr. Jno. Hammond; s. Davyson 7th 476; d. Davy 27th 1451 by Davyson 5th 287; gr d. Davy 5th 167 by Tenant Farmer 213. See Davy—H 1.

2132   DAVY 39TH — H 1.

Calved September 16, 1881; breeder Mr. Jno. Hammond; s. Davyson 7th 476; d. Davy 15th 844 by Davyson 3rd 48; gr d. Davy 5th 167 by Tenant Farmer 213. See Davy—H 1.

2133   DAVY 41ST — H 1.

Calved January 19, 1882; breeder Mr. Jno. Hammond; s. Davyson 7th 476; d. Davy 5th 167 by Tenant Farmer 213; gr d. Davy—H 1.

REGISTER OF COWS.

2134   DAVY 42ND — H 1.

Calved June 6, 1882; breeder Mr. Jno. Hammond; s. Davyson 7th 476; d. Davy 21st 1445 by Davyson 5th 287; gr d. Davy 7th 169 by Young Duke 234. See Davy 2nd—H 1 by Sir Nicholas 202.

2135   DAVY 43RD — H 1.

Calved August 25, 1882; breeder Mr. Jno. Hammond; s. Davyson 7th 476; d. Davy 22nd 1446 by Davyson 5th 287; gr d. Davy 16th 845 by Redjacket 7th 169. See Davy 7th—H 1 by Young Duke 234.

2136   DAVY 44TH — H 1.

Calved August 16, 1882; breeder Mr. Jno. Hammond; s. Davyson 7th 476; d. Davy 27th 1451 by Davyson 5th 287; gr d. Davy 5th 167 by Tenant Farmer 213. See Davy—H 1.

2137   DAVY 45TH — H 1.

Calved October 2, 1882; breeder Mr. Jno. Hammond; s. Davyson 7th 476; d. Davy 28th 1452 by Davyson 6th 475; gr d. Davy 15th 844 by Davyson 3rd 48. See Davy 5th—H 1 by Tenant Farmer 213.

2138   DAVY 46TH — H 1.

Calved January 15, 1882; breeder Mr. Jno. Hammond; s. Davyson 6th 475; d. Davy 29th 1453 by Davyson 6th 475; gr d. Davy 7th 169 by Young Duke 234. See Davy 2nd—H 1 by Sir Nicholas 202.

2139   DAVY 47TH — H 1.

Calved May 24, 1882; breeder Mr. Jno. Hammond; s. Davy Butler 483; d. Davy 23rd 1447 by Davyson 5th 287; gr d. Davy 17th 846 by Redjacket 7th 169. See Davy 4th—H 1 by Tenant Farmer 213.

2140   DAVY 48TH — H 1.

Calved May 27, 1882; breeder Mr. Jno. Hammond; s. Davy Butler 483; d. Davy 34th 1458 by Davyson 6th 475; gr d. Davy 15th 844 by Davyson 3rd 48. See Davy 5th—H 1 by Tenant Farmer 213.

2141   DAVY 49TH — H 1.

Calved July 1, 1882; breeder Mr. Jno. Hammond; s. Davy Butler 483; d. Davy 35th 1459 by Davyson 6th 475; gr d. Davy 5th 167 by Tenant Farmer 213. See Davy—H 1.

2142   DAVY 50TH — H 1.

Calved January 6, 1883; breeder Mr. Jno. Hammond; s. Davy Butler 483; d. Davy 29th 1453 by Davyson 6th 475; gr d. Davy 7th 169 by Young Duke 234. See Davy 2nd—H 1 by Sir Nicholas 202.

2143   DAVY 51ST — H 1.

Calved January, 1883; breeder Mr. Jno. Hammond; s. Davy Butler 483; d. Davy 5th 169 by Tenant Farmer 213; gr d. Davy—H 1.

2144   DAVY DUCHESS 2ND — H 1.

Calved March 19, 1881; breeder the Right Hon. Lord Hastings; s. Thornham Duke 2nd 585; d. Davy 16th 845 by Redjacket 7th 169; gr d. Davy 7th 169 by Young Duke 234. See Davy 2nd—H 1 by Tenant Farmer 213.

2145   DAVY DUCHESS 3RD — H 1.

Calved April 20, 1882; breeder the Right Hon. Lord Hastings; s. Davyson 7th 476; d. Davy 16th 845 by Redjacket 7th 169; gr d. Davy 7th 169 by Young Duke 234. See Davy 2nd—H 1 by Sir Nicholas 202.

2146   DAVY PRINCESS — H 1.

[Davy 40th]—Calved August, 1881; breeder Mr. Jno. Hammond; owner Col. J. B. Mead and Mr. R. J. Kimball; s. Davyson 7th 476; d. Davy 20th 1444 by Davyson 4th 286; gr d. Davy 5th 167 by Tenant Farmer 213. See Davy—H 1.

2147   DIANA — AA 1.

Calved March 24, 1882; breeder Mr. G. F. Taber; owner Mr. D. L. Stevens; s. Champion 271; d. Lucilla 1009—AA 1 by Ravinewood Beau 160; gr d. Ravinewood Lass 455 by Robin 176. See Nelly—A 1 by Hero 2nd 86.

2148   DIDLINGTON DAVY — H 1.

Calved January 27, 1882; breeder Mr. W. A. T. Amherst, M.P.; s. Davyson 7th 476; d. Davy 24th 1448 by Davyson 5th 287; gr d. Davy 15th 844 by Davyson 3rd 48. See Davy 5th—H 1 by Tenant Farmer 213.

2149   DIDO — S 3.

Calved March 9, 1882; breeder Mr. Garrett Taylor; s. Davyson 8th 477; d. Dora 854 by Norfolk Duke 127; gr d. Dorothy 182 by George of Elmham 76. See Stoke—S 3 by Elmham 65.

REGISTER OF COWS.

2150   DIDO − 1 SUFF.
Calved November, 1879; breeder Mr. Crane, Sweffling; owner Mr. C. Austin; s. Gamester 310; d. Baker 1992 by late Mr. C. Austin's Bull.

2151   DINGOO − R 8.
Calved November, 1874; breeder Mr. T. Easter; owner Mr. Garrett Taylor; s. Read 385; d. Beauty—R 8 by a Laxfield Bull.

2152   DORA − P 9.
Calved July 22, 1881; breeder Mr. W. A. T. Amherst, M.P.; s. Norfolk John 2nd 527; d. Dolly 1464 by Norfolk John 131; gr d. Daisy 1436 by Redjacket 169. See Cherry—P 9.

2153   DORCAS − R 8.
Calved January, 1882; breeder Mr. T. Easter; owner Mr. Garrett Taylor; s. Brundish Prince 462; d. Dingoo 2151 by Read 385; gr d. Beauty—R 8 by a Laxfield Bull.

2154 DUCHESS OF HAMILTON − A 4.
Calved October 13, 1880; breeder Her Grace the Duchess of Hamilton; owner Col. J. B. Mead and Mr. R. J. Kimball; s. Handsome Prince 317; d. Little Katie 1630 by Royal Duke 181; gr d. Kattie 975 by Benedict 17. See Ringlet 2nd—A 4 by Tenant Farmer 213.

2155 DUCHESS OF SUFFOLK 2ND − O 1.
Calved November 1882; breeder Sir E. C. Kerrison, Bart.; s. Lofty 515; d. Oakley 4th by Roundhead 400; gr d. Oakley 3rd 400 by Major 109. See Oakley—O 1 by Rifleman 175.

2156   DUMMY − O 9.
Calved February 23, 1882; breeder Mr. Garrett Taylor; s. Cato 468; d. Silent Girl 1854 by Osman 530; gr d. Silent Lass 1189 by Powell 143. See Silence—O 9 by Rifleman 185.

2157   DUTCH OVEN − 1 SUFF.
Calved May 27, 1872; breeder Mr. C. Austin; s. Shylock 571; d. Dido 2150 by Gamester 310; gr d. Baker 1992 by late Mr. C. Austin's Bull.

REGISTER OF COWS.

2158   EDITH — N 6.
Calved October, 1876; breeder Mr. W. Bradfield; owner Mr. H. Haylock; s. Rufus 188; d. Cherry 94 by Fransham Captain 71; gr d. Tit—N 6 by Necton 3rd 122.

2159   ELIZABETH — E 13.
Calved October 12, 1881; breeder Mr. H. Birkbeck; s. Haman 499; d. Elmham Taylor 1493 by Rufus 188; gr d. Cheerful 761 by Cringleford Duke 43. See Barker—E 13.

2160   ELMHAM NELLY 3RD — A 1.
Calved July 28, 1881; breeder Mr. R. E. Lofft; s. Hector 319; d. Elmham Nelly 2nd 1492 by Roland 397; gr d. [Elmham] Nelly 371 by Hero 2nd 86. See Primrose—A 1 by Elmham Sire 67.

2161   ELMHAM ROSEBUD 3RD — A 1.
Calved July 15, 1880; breeder Mr. R. E. Lofft; s. Hector 319; d. Elmham Rosebud 2nd 872 by Prince Regent 381; gr d. [Elmham] Rosebud 195 by Hero 2nd 86. See Rose—A 1 by Redjacket 2nd 164.

2162   ELMHAM ROSEBUD 4TH — A 1.
Calved May 3, 1882; breeder Mr. R. E. Lofft; s. Rinaldo 556; d. Elmham Rosebud 3rd 2161 by Hector 319; gr d. Elmham Rosebud 2nd 872 by Prince Regent 381. See [Elmham] Rosebud—A 1 by Hero 2nd 86.

2163   ELVIRA — A 33.
Calved May, 1881; breeder Mr. R. H. Mason; s. Slasher 577; d. Elm-leaf 1488 by a Ramsley Bull; gr d. a Ramsley Cow—A 33.

2164   EMMA OF ELMHAM — L 3.
Calved March 31, 1882; breeder Mr. T. Fulcher; s. Lofty 515; d. Elm-branch 1482 by Rufus 188; gr d. Elmer—L 3 by Elmham Sire 67.

2165   EMMELINE — W 14.
Calved February, 1880; breeder Her Grace the Duchess of Hamilton; owner Mr. G. F. Taber; s. Handsome Prince 317; d. Esmeralda 873 by Roundhead 180; gr d. Emerald 204 by Stoke Duke 209. See Clara—W 14.

2166   ENCHANTRESS — W 14.

Calved October 20, 1882; breeder Mr. F. D. Kent; s. Young Prince 608; d. Witch 657 by Tommy 216; gr d. Clara—W 14.

2167   ERA — I 1.

Calved December 6, 1880; breeder Mr. T. Brown; s. Priam 373; d. Ethel 876 by The Peer 139; gr d. Beauty 16th 32 by Tenant Farmer 213. See Beauty 3rd—I 1 by an Elmham Bull.

2168   ERICA — I 1.

Calved October 25, 1881; breeder Mr. T. Brown; s. Priam 373; d. Ethel 876 by The Peer 139; gr d. Beauty 16th 32 by Tenant Farmer 213. See Beauty 3rd—I 1 by an Elmham Bull.

2169   ESTHER 2ND — E 3.

Calved March 25, 1882; breeder Mr. J. F. Rogers; s. Emperor 489; d. Esther 874 by Nicholson 360; gr d. Eaton Beryl 864 by Powell 148. See Cherry—E 3 by Stoke 208.

2170   EUGENIE — E 12.

Calved January 4, 1882; breeder Mr. J. F. Rogers; s. Emperor 489; d. Empress 1495 by Suffolk 211; gr d. Susanna 587 by Stoke Duke 209. See Susan—E 12.

2171   EULALIE — N 4.

Calved August, 1882; breeder Mr. R. H. Mason; s. Philip 538; d. Eugenie 1499 by Popgun 542; gr d. Empress 1496 by King Harry 332. See Rose 2nd—N 4 by Longham 104.

2172   EUROTAS — E 12.

Calved November 26, 1882; breeder Mr. J. F. Rogers; s. Emperor 489; d. Empress 1495 by Suffolk 211; gr d. Susanna 587 by Stoke Duke 209. See Susan—E 12.

2173   EYEBRIGHT 3RD — O 6.

Calved December, 1880; breeder Sir E. C. Kerrison, Bart.; s. Dick's Son 482; d. Eyebright 2nd 879 by Harold 83; gr d. Mirth 344 by Rifleman 175. See Vanity—O 6.

REGISTER OF COWS.

2174   EYKE BELLE – B 11.
Calved 1880; breeder Mr. A. J. Smith; s. Pickwick 720; d. Belle 2010 by Iron Duke 125; gr d. Bellona 705 by The Baron 10. See Suffolk Belle—B 11 by Seneca 195.

2175   EYKE DUCHESS – B 11.
Calved December, 1880; breeder Mr. A. J. Smith; s. Pickwick 720; d. Suffolk Duchess 2557 by Crown Prince 281; gr d. Suffolk—B 11.

2176   EYKE GIPSY – B 19.
Calved 1880; breeder Mr. A. J. Smith; s. Pickwick 720; d. Fortune Teller 903 by The Baron 10; gr d. Gipsy—B 19.

2177   EYKE LASSIE – B 11.
Calved 1879; breeder Mr. A. J. Smith; s. Pickwick 720; d. Playford Lassie 1077 by The Baron 10; gr d. Suffolk Belle 582 by Seneca 195. See Suffolk—B 11.

2178   FAIRY 2ND – F 8.
Calved December 6, 1881; breeder Mr. J. F. Rogers; s. Emperor 489; d. Fairy 881 by Suffolk 211; gr d. Fanny—F 8.

2179   FAIRY – V 3.
Calved April 27, 1882; breeder Mr. G. Gooderham; s. Wild Roger 603; d. Fancy 1508 by Gamester 310; gr d. Fay 1511 by Troston 424. See Dainty—V 3 by Perfection 140.

2181   FAME – A 11.
Calved November 14, 1881; breeder Rev. A. G. Legge; owner Mr. T. Fulcher; s. Lofty 515; d. Frolic 906 by The Palmer 138; gr d. Fanny Bradfield 891 by Money 352. See Nancy—A 11.

2182   FAN OF ELMHAM – A 11.
Calved November 24, 1882; breeder Rev. A. G. Legge; s. Tommy 588; d. Frances 904 by Harry 84; gr d. Fanny Bradfield 891 by Money 352. See Nancy—A 11.

2183   FANCY — R 8.

Calved June 14, 1882; breeder Mr. W. B. Easter; owner Mr. J. Boggis, Jun.; s. Brundish Prince 462; d. Fancy 1507 by Simon 408; gr d. Beauty 696 by Read 385. See Beauty—R 8 by a Laxfield Bull.

2184   FANCY SPOT — W 9.

Calved October 26, 1881; breeder the Most Hon. the Marquis of Bristol; s. Fancy King 491; d. Lady Spotless 304 by Cremorne 42; gr d. Lady Spot 303 by Seneca 195. See Lady—W 9.

2185   FANNIE — O 6.

Calved April 12, 1881; breeder Sir J. W. C. Hartopp, Bart.; s. Hardwick 501; d. Canteen 1361 by a Son of Rifleman 175; gr d. Mirth 344 by Rifleman 175. See Vanity—O 6.

2186   FATIMA — P 3.

Calved February 1, 1882; breeder Mr. T. Brown; s. Priam 373; d. Faith 1504 by Norfolk Duke 127; gr d. Florence 898 by The Palmer 138. See Thursford Rose—P 3 by Norfolk Duke 127.

2187   FILLPAIL — A 24.

Calved June 24, 1880; breeder the Right Hon. Lord Suffield, K.C.B.; s. Norfolk 361; d. Flower 901 by Rufus 187; gr d. Bridget 723 by Witton 432. See Floss—A 24 by Witton 432.

2188   FILLPAIL — S 2.

Calved 1878; breeder Mr. H. Birkbeck; owner Sir J. W. C. Hartopp, Bart.; s. Roundhead 180; d. Harebell 254 by Powell 143; gr d. Holly 270 by Tommy 216. See Heartsease—S 2 by Tommy 216.

2189   FINE FRUIT — B 18.

Calved 1879; breeder Mr. A. J. Smith; s. Pickwick 720; d. Fancy Fruit 887 by The Baron 10; gr d. Fancy—B 18.

2190   FLAME — F 8.

Calved November 7, 1881; breeder Mr. J. F. Rogers; s. Emperor 489; d. Fanny—F 8.

2191　　　　　FLAME — P 3.

Calved November 28, 1881; breeder Mr. T. Brown; s. Priam 373; d. Fusee 1530 by Royal Duke 181; gr d. Florence 898 by The Palmer 138. See Thursford Rose—P 3 by Norfolk Duke 127.

2192　　　　　FLECK — N 4.

Calved March 20, 1881; breeder Mr. B. M. Haggard; s. Young King Cole 607; d. Marguerite 1657 by King Cole 330; gr d. Daisy 2nd by Prince Charlie 151. See Daisy—N 4 by Necton 3rd 122.

2193　　　　　FLO — K 24.

Calved November, 1882; breeder Mr. W. Hudson; s. The Doctor 486; d. Flora 1520 by Davyson 3rd 48; gr d. Sal 2nd by Wonder 251. See Sal—K 24.

2194　　　　　FLORA — S 2.

Calved 1879; breeder Sir J. W. C. Hartopp, Bart.; s. Bounty 460; d. Red-berry 1765 by Roundhead 180; gr d. Harebell 254 by Powell 143. See Holly—S 2 by Tommy 216.

2195　　　　　FLORENCE — A 24.

Calved November 5, 1879; breeder the Right Hon. Lord Suffield, K.C.B.; s. Norfolk 361; d. Floss 2nd 1522 by Rufus 187; gr d. Floss 899 by Witton 432. See Floss—A 24.

2196　　　　　FLOSSIE — U 2.

Calved July 23, 1881; breeder Mr. R. E. Lofft; owners Col. J. B. Mead and Mr. R. J. Kimball; s. Stout 581; d. Dolly 179 by Sampson 171; gr d. Floss—U 2.

2197　　　　　FRAGRANCE — P 3.

Calved June 14, 1881; breeder Mr. T. Brown; s. Priam 373; d. Florence 898 by The Palmer 138; gr d. Thursford Rose 600 by Norfolk Duke 127. See Rose—P 3.

2198　　　　　FRANCES — N 4.

Calved November, 1880; breeder Mr. R. H. Mason; s. Slasher 577; d. Fransham by King Harry 332; gr d. Daisy 3rd 1434 by Bradfield 264. See Daisy—N 4 by Necton 3rd 122.

2199   FREDA — A 23.

Calved November 4, 1879; breeder the Right Hon. Lord Suffield, K.C.B.; s. Norfolk 361; d. Winifred 1273 by Witton 432; gr d. Dido—A 23 by Hero of Newcastle 85.

2200   FRESH FRUIT — B 18.

Calved 1881; breeder Mr. A. J. Smith; s. Pickwick 720; d. Fancy Fruit 887 by The Baron 10; gr d. Fancy—B 18.

2201   FRIDAY 4TH — T 17.

Calved February 28, 1882; breeder Mr. John Howell; s. Davyson 9th 478; d. Friday 905 by Masker 346; gr d. Abbess—T 17.

2202   FRIDAY 5TH — T 17.

Calved March 9, 1882; breeder Mr. John Howell; s. Davyson 9th 478; d. Friday 2nd 1526 by Sir Thomas 409; gr d. Friday 905 by Masker 346. See Abbess—T 17.

2203   FRIDAY 6TH — T 17.

Calved January 15, 1883; breeder Mr. John Howell; s. Davyson 9th 478; d. Friday 2nd 1526 by Sir Thomas 409; gr d. Friday 905 by Masker 346. See Abbess—T 17.

2204   FUCHSIA — V 1.

Calved September 12, 1882; breeder Mr. W. A. T. Amherst, M.P.; s. Davyson 3rd 48; d. Flirt 2nd 1516 by Troston 424; gr d. Flirt 895 by Councillor 38. See Rosebud—V 1 by Doncaster 50.

2205   GLAD — V 10.

Calved January, 1882; breeder Mr. J. M. Spinks; s. Lord George 520; d. Gadfly 1532 by Damian 282; gr d. Grimace 3rd 664 by Monarch 241. See Grimace 2nd—V 10 by Bullfinch 239.

2206   GAINFUL — V 10.

Calved May 1, 1882; breeder Mr. J. M. Spinks; s. Lord George 520; d. Gain 1533 by Max 112 [incorrectly registered in 1881 as by Damian 282]; gr d. Gadfly 1532 by Damian 282. See Grimace 3rd—V 10 by Monarch 241.

2207   GARNET — V 13.

Calved February, 1882; breeder Mr. J. M. Spinks; s. Lord George 520; d. Grace 925 by Rendham Wonder 245; gr d. Lady Rowley 985 by Monarch 241. See Rowley—V 13 by a Glemham Bull.

2208   GEM, THE — W 14.

Calved February 10, 1882; breeder Her Grace the Duchess of Hamilton; s. Suffolk Baronet 583; d. The Easton Gem 868 by Baron Handsome 254; gr d. Emerald 204 by Stoke Duke 209. See Clara—W 14.

2209   GENTLE ROSY — B 9.

Calved January 18, 1882; breeder Mr. W. A. T. Amherst, M.P.; s. Shylock 572; d. Gentle Rose 914 by Iron Duke 125; gr d. Dwarf Rose 193 by Cremorne 42. See Rose—B 9.

2210   GIDDY GAL — K 15.

Calved April 21, 1881; breeder Mr. Garrett Taylor; s. Grey Spot 498; d. Flirt 893 by Roundhead 180; gr d. Fairy 882 by Lord Easton 105. See Fanciful—K 15 by Cherry Duke 32.

2212   GLAZE — V 9.

Calved December, 1880; breeder Mr. J. M. Spinks; owner Mr. Garrett Taylor; s. Lord George 520; d. Glad 920 by Max 112; gr d. Glee 2nd 663 by Monarch 241. See Glee—V 9 by Bullfinch 239.

2213   GLEANER — V 9.

Calved December, 1882; breeder Mr. J. M. Spinks; s. Lord George 520; d. Glad 920 by Max 112; gr d. Glee 2nd 663 by Monarch 241. See Glee—V 9 by Bullfinch 239.

2214   GLIDE — V 9.

Calved August, 1882; breeder Mr. J. M. Spinks; s. Lord George 520; d. Glen 922 by Max 112; gr d. Glee 2nd 663 by Monarch 241. See Glee—V 9 by Bullfinch 239.

2215   GLIMMER — V 13.

Calved January, 1881; breeder Mr. J. M. Spinks; s. Lord George 520; d. Glemham 923 by Max 112; gr d. Lady Rowley 985 by Monarch 241. See Rowley—V 13 by a Glemham Bull.

2216   GLOSS — P 4.

Calved 1881; breeder Sir J. W. C. Hartopp, Bart.; s. Bounty 460; d. Milkmaid 2367 by Roundhead 180; gr d. Nina 5th 1053 by Norfolk Duke 127. See Nina 2nd—P 4 by Tenant Farmer 213.

2217   GLOSS 5TH — V 11.

Calved February 28, 1881; breeder Mr. R. E. Lofft; s. Long 516; d. Gloss 2nd 665 by Boss 237; gr d. Gloss—V 11 by a Glemham Bull.

2218   GLOSS 6TH — V 11.

Calved August 22, 1881; breeder Mr. R. E. Lofft; s. Bantam 451; d. Gloss 5th 2217 by Long 516; gr d. Gloss 2nd 665 by Boss 237. See Gloss—V 11 by a Glemham Bull.

2219   GLUNBERRY — E 11.

Calved June 1, 1881; breeder Sir J. W. C. Hartopp, Bart.; s. Hardwick 501; d. Gladiolus 1537 by Brutus 269; gr d. Pansy 1063 by Cringleford Duke 43. See Pretty—E 11 by Cantley 29.

2220   GOLDIE — I 13.

Calved November 25, 1881; breeder Mr. R. E. Lofft; owner the Right Hon. W. H. Smith, M.P.; s. Bantam 451; d. Rosebud 6th 1801 by Bright 267; gr d. Rosebud 2nd 1153 by Rudham Hero 183. See Rosebud—I 13.

2221   GREENFIELD — A 5.

Calved 1881; breeder Sir J. W. C. Hartopp, Bart.; s. Hardwick 501; d. Mrs. Bradfield 1686 by Rufus 188; gr d. Ramsley 2nd 1123 by The Palmer 138. See Ramsley—A 5 by Hero of Newcastle 85.

2222   GRESSENHALL — 3 NORF.

Calved 1878; breeder Mr. J. Nicholson; owner Mr. Jno. Baly; s. an Elmham Bull; d. an Elmham Cow. See Introduction.

2223   GRIM — V 10.

Calved May, 1881; breeder Mr. J. M. Spinks; s. Lord Charles 693; d. Grimace 7th by Lord George 520; gr d. Grimace 4th 927 by Rendham Wonder 245. See Grimace 3rd—V 10 by Monarch 241.

REGISTER OF COWS.

2224 GUELDER ROSE – B 9.

Calved January 4, 1882; breeder Mr. H. Biddell; s. Monarch 4th 351; d. Standard Rose 1867 by Crown Prince 281; gr d. Rose—B 9.

2225 GLEMHAM ROSE 2ND – V 14.

Calved August, 1880; breeder Her Grace the Duchess of Hamilton; owner Mr. G. F. Taber; s. Handsome Prince 317; d. Glemham Rose 921 by Young Monarch 246; gr d. Fillpail by Duke 239. See Honesty—V 14.

2226 HAIRBELL – M 2.

Calved December 3, 1880; breeder Mr. T. Brown; s. Priam 373; d. Hortensia 953 by Norfolk Duke 127; gr d. Rose 4th 474 by Tenant Farmer 213. See Red Rose—M 2.

2227 HANDMAID – M 2.

Calved March 15, 1881; breeder Mr. T. Brown; s. Favourite 492; d. Hetty 268 by The Peer 139; gr d. Rose 5th 475 by Tenant Farmer 213. See Red Rose—M 2.

2228 HANDSOME – I 16.

Calved 1879: breeder Mr. W. Hudson; s. Davyson 5th 287; d. Countess 1406 by Quarles Duke 548; gr d. Coral—I 16 by Proud 547.

2229 HANDSOME 2ND – T 17.

Calved October 8, 1881; breeder Mr. John Howell; s. Davyson 9th 478; d. Handsome 1553 by Sir Thomas 409; gr d. Strawberry 1874 by Masker 346. See Friday—T 17 by Masker 346.

2230 HANDSOME 11TH – U 3.

Calved September 27, 1880; breeder Mr. R. E. Lofft; s. Hector 319; d. Handsome 5th 935 by Troston Hero 221; gr d. Handsome 2nd 249 by Sampson 191. See Handsome—U 3.

2231 HANDSOME 12TH – U 3.

Calved February 1, 1881; breeder Mr. R. E. Lofft; s. Hector 319; d. Handsome 6th 936 by Cherry Duke 32; gr d. Handsome 2nd 249 by Sampson 191. See Handsome—U 3.

REGISTER OF COWS.

2232   HANDSOME 13TH – U 3.
Calved December 5, 1881; breeder Mr. R. E. Lofft; s. Rinaldo 556; d. Handsome 9th 1555 by Pryor 609; gr d. Handsome 8th 1554 by Bright 267. See Handsome 5th—U 3 by Troston Hero 221.

2233   HANDSOME 14TH – U 3.
Calved February 26, 1882; breeder Mr. R. E. Lofft; s. Doubtful 487; d. Handsome 6th 936 by Cherry Duke 82; gr d. Handsome 2nd 249 by Sampson 191. See Handsome—U 3.

2234   HANDSOME 15TH – U 3.
Calved May 15, 1882; breeder Mr. R. E. Lofft; s. Long 516; d. Handsome 10th 1556 by Hector 319; gr d. Handsome 3rd 250 by Sampson 191. See Handsome—U 3.

2235
HANDSOME OF TITTLESHALL – 1 NORF.
Calved 1877; breeder Mr. W. Wiffin; owner Mr. John Baly; s. an Elmham Bull; d. a Pond Cow. See Introduction.

2236
HANDSOME OF GRESSENHALL – 3 NORF.
Calved 1878; breeder Mr. James Nicholson; owner Mr. John Baly; s. an Elmham Bull; d. an Elmham Cow. See Introduction.

2237   HANDSOME – 2 SUFF.
Calved 1879; breeder Mr. E. Boon; s. a Wolton Bull; d. Cossett by an Oakley Bull; gr d. Nancy by a Kettleburgh (Turner) Bull. See Introduction.

2238   HANDSOME ROSE – O 14.
Calved July 5, 1881; breeder the Right Hon. Lord Henniker; owner Mr. G. F. Taber; s. Cyprus 473; d. Roseleaf 1159 by Ruddy 492; gr d. Cherry—O 14.

2239   HANDSOME RUBY – O 2.
Calved March 22, 1881; breeder Her Grace the Duchess of Hamilton; owner Mr. G. F. Taber; s. Monk 525; d. Ruby 2nd 1824 by Marquis 344; gr d. Ruby 1164 by Marquis 111. See Queen 2nd—O 2 by Major 109.

2240   HANNAH — P 1.
Calved March 7, 1882; breeder Mr. Garrett Taylor; s. Cato 468; d. Hecate 1564 by Rufus 188; gr d. Lydia 2nd 1011 by Powell 143. See Handsome 3rd—P 1 by Tenant Farmer 213.

2241   HAREBELL — X 5.
Calved March, 1879; breeder Her Grace the Duchess of Hamilton; s. Handsome Prince 317; d. Blue Bell—X 5 by Youngster 439.

2242   HARP — M 2.
Calved December 3, 1880; breeder Mr. T. Brown; s. Priam 373; d. Helen 265 by The Peer 139; gr d. Rose 4th 474 by Tenant Farmer 213. See Red Rose—M 2.

2243   HARRIETT — S 2.
Calved October 20, 1881; breeder Mr. H. Birkbeck; s. Haman 499; d. Harebell 254 by Powell 143; gr d. Holly 270 by Tommy 216. See Heartsease—S 2 by Tommy 216.

2244   HEARTSEASE — O 11.
Calved October 7, 1882; breeder Mr. Garrett Taylor; s. Cyprus 473; d. Pansy 1722 by Crown Prince 281; gr d. Thornham Polly 1229 by Eclipse 63. See Polly—O 11.

2245   HEATH — M 2.
Calved January 30, 1881; breeder Mr. T. Brown; s. Priam 373; d. Hester 267 by The Peer 139; gr d. Rose 5th 475 by Tenant Farmer 213. See Red Rose—M 2.

2246   HELEN — A 1.
Calved January 30, 1882; breeder Mr. H. Haylock; s. The Parson 533; d. Blossom 2023 by Rufus 188; gr d. Nellie 1702 by The Palmer 138. See Nelly—A 1 by Hero 2nd 86.

2247   HELENA 3RD — M 2.
Calved July 24, 1881; breeder Mr. R. E. Lofft; s. Doubtful 487; d. Helena 944 by Benedict 17; gr d. Rose 4th 474 by Tenant Farmer 213. See Red Rose—M 2.

2248 HEMP — M 2.

Calved July 27, 1881; breeder Mr. T. Brown; s. Peter Parley 537; d. Hawthorn 1561 by Royal Duke 181; gr d. Helen 265 by The Peer 139. See Rose 4th—M 2 by Tenant Farmer 213.

2249 HERMIA — M 2.

Calved February 1, 1882; breeder Mr. T. Brown; s. Priam 373; d. Hester 267 by The Peer 139; gr d. Rose 5th 475 by Tenant Farmer 213. See Red Rose—M 2.

2250 HERMIONE — M 2.

Calved November 27, 1881; breeder Mr. T. Brown; s. Priam 373; d. Hortensia 953 by Norfolk Duke 127; gr d. Rose 4th 474 by Tenant Farmer 213. See Red Rose—M 2.

2251 HEROINE — N 6.

Calved April, 1882; breeder Mr. R. H. Mason; s. Philip 538; d. Darling 1443 by King Cole 330; gr d. Violet 4th 1252 by Lord Easton 105. See Dainty—N 6 by Prince Charlie 151.

2252 HESTER—2 SUFF.

Calved July, 1882; breeder Mr. E. Boon; s. Wild Roger 603; d. Handsome 2237 by a Wolton Bull; gr d. Cossett by an Oakley Bull.

2253 HILDA — W 3.

Calved July 2, 1882; breeder Mr. W. A. T. Amherst, M.P.; s. Davyson 3rd 48; d. Helene 945 by Powell 143; gr d. Helen 266 by Norfolk Duke 127. See Nelly of Newbourn—W 3 by Prince Regent 153.

2254 HONEY-DEW — M 2.

Calved July 12, 1882; breeder Mr. T. Brown; s. Priam 373; d. Hawthorn 1561 by Royal Duke 181; gr d. Helen 265 by The Peer 139. See Rose 4th—M 2 by Tenant Farmer 213.

2255 HONEY-DEW — V 17.

Calved July 22, 1882; breeder Col. W. B. Long; s. Butley 465; d. Honey 950 by Nero 358; gr d. Una by Prince 375. See Cherry—V 17 by Hero 322.

2256   HONEY-WOOD — B 12.

Calved February 1, 1882; breeder Her Grace the Duchess of Hamilton; s. Suffolk Baronet 583; d. Little Bee 1003 by Crown Prince 281; gr d. Queen Bee 450 by Seneca 195.   See Bee—B 12.

2257   HOSTESS — M 2.

Calved August 22, 1882; breeder Mr. T. Brown; s. Priam 373; d. Helen 265 by The Peer 139; gr d. Rose 4th 474 by Tenant Farmer 213.   See Red Rose—M 2.

2258   HYPATIA — M 2.

Calved October 28, 1881; breeder Mr. T. Brown; s. Priam 373; d. Helen 265 by The Peer 139; gr d. Rose 4th 474 by Tenant Farmer 213.   See Red Rose—M 2.

2259   IRENE — AA 1.

Calved March 13, 1882; breeder Mr. G. F. Taber; s. Champion 271; d. Lida 1620 by Champion 271; gr d. Lucilla 1009—AA 1 by Ravinewood Beau 160. See Ravinewood Lass—A 1 by Robin 176.

2260   IRIS — V 2.

Calved December 17, 1881; breeder Mr. C. Austin; s. Wild Robin 600; d. Flora 2nd 897 by Doncaster 50; gr d. Flora 229 by King Alfred 96. See Red Stockings—V 2 by Wonder 230.

2261   ISABEL — I 17.

Calved July, 1881; breeder Mr. W. Hudson; s. The Doctor 486; d. Isabella 2nd 1590 by Davyson 5th 287; gr d. Isabella 1589 by Proud 547. See Isabella—I 17.

2262   IVY — A 33.

Calved April 23, 1882; breeder Mrs. E. Perkins; s. Othello 532; d. Elm-leaf 3rd 1490 by King Cole 330; gr d. Elm-leaf—A 33 by a Ramsley Bull.

2263   IVY — P 2.

Calved January 31, 1882; breeder Mr. Garrett Taylor; s. Cato 468; d. Isabel 1588 by Duke of Norfolk 295; gr d. Isabelle 278 by Young Duke 234. See Strawberry 2nd—P 2 by Tenant Farmer 213.

2264   JEMIMA − 1 Norf.

Calved June, 1882; breeder Mr. Rivett; s. Falstaff 303; d. Jenny 2266 by Falstaff 303; gr d. Lucy 2338 by an Elmham Bull.

2265   JENNIE − A 13.

Calved September 30, 1881; breeder Sir J. W. C. Hartopp, Bart.; s. Hardwick 501; d. Red Stocking 1778 by Rufus 188; gr d. Spot—A 13.

2266   JENNY − 1 Norf.

Calved 1878; breeder Mr. J. Rivett; s. Falstaff 303; d. Lucy 2338 by an Elmham Bull; gr d. a Pond Cow.

2267   JESSICA − R 1.

Calved May 21, 1882; breeder Mr. H. Biddell; s. Shylock 572; d. Susanna 1219 by Norfolk Duke 127; gr d. Susan 586 by Tommy 216. See Sarah—R 1 by Elmham 65.

2268   JESSIE − A 2.

Calved December 17, 1880; breeder the Right Hon. Lord Suffield, K.C.B.; s. Norfolk 361; d. Judith 1601 by Rufus 187; gr d. Judy 966 by Witton 432. See Daffodil—A 2 by Hero of Newcastle 85.

2269   JESSIE − A 5.

Calved 1880; breeder Sir J. W. C. Hartopp, Bart.; s. Bounty 460; d. Mrs. Bradfield 1686 by Rufus 188; gr d. Ramsley 2nd 1123 by The Palmer 138. See Ramsley—A 5 by Hero of Newcastle 85.

2270   JESSIE − I 21.

Calved 1879; breeder Mr. W. Hudson; s. Davyson 5th 287; d. Sepsy 1840 by Proud 547; gr d. Sepsy—I 21 by Redjacket 6th 168.

2271   JET − I 2.

Calved August 8, 1881; breeder Mr. T. Brown; s. Peter Parley 537; d. Jessamine 1594 by Royal Duke 181; gr d. Jane 957 by The Peer 139. See Ruby 4th—I 2 by Tenant Farmer 213.

REGISTER OF COWS.

2272 JEWEL 2ND — O 8.

Calved January, 1881; breeder Sir E. C. Kerrison, Bart.; s. Lofty 515; d. Jewel 281 by Rufus 3rd 186; gr d. Mary Grey—O 8.

2273 JEWESS — O 8.

Calved September, 1882; breeder Sir E. C. Kerrison, Bart.; s. Lofty 515; d. Jewel 281 by Rufus 3rd 186; gr d. Mary Grey—O 8.

2274 JILT — V 14.

Calved April 13, 1881; breeder Her Grace the Duchess of Hamilton; owner General L. F. Ross; s. Handsome Prince 317; d. Rosebud 1156 by Baron Handsome 254; gr d. Glemham Rose 921 by Young Monarch 246. See Fillpail—V 14 by Duke 293.

2275 JONQUIL — I 2.

Calved April 14, 1881; breeder Mr. T. Brown; s. Priam 373; d. Jenny 270 by The Peer 139; gr d. Ruby 4th 518 by Tenant Farmer 213. See Ruby 2nd—I 2 by Hero 2nd 86.

2276 JOSEPHINE — I 2.

Calved August 3, 1881; breeder Mr. T. Brown; s. Priam 373; d. Jessica 1595 by Royal Duke 181; gr d. Jenny 279 by The Peer 139. See Ruby 4th—I 2 by Tenant Farmer 213.

2277 JOYCE — I 2.

Calved June 11, 1882; breeder Mr. T. Brown; s. Priam 373; d. Jane 957 by The Peer 139; gr d. Ruby 4th 518 by Tenant Farmer 213. See Ruby 2nd—I 2 by Hero 2nd 86.

2278 JUDY 2ND — N 6.

Calved August 20, 1882; breeder Mr. B. M. Haggard; s. Anthony 448; d. Tit 3rd 1898 by King Cole 330; gr d. Tit 2nd 1897 by Lord Easton 105. See Tit—N 6 by Necton 3rd 122.

2279 JUICE — O 8.

Calved July 2, 1882; breeder Mr. A. Taylor; s. Starston Duke 570; d. Jewel 281 by Rufus 3rd 186; gr d. Mary Grey—O 8.

REGISTER OF COWS.

2280   JULIA — I 2.

Calved June 13, 1882; breeder Mr. T. Brown; s. Priam 373; d. Jessica 1595 by Royal Duke 181; gr d. Jenny 279 by The Peer 139. See Ruby 4th—I 2 by Tenant Farmer 213.

2281   JULIET — U 23.

Calved 1881; breeder Mr. W. Harvey; s. Hamlet 500; d. Olga. See Annie—U 23 by a Red Bull.

2282   JUNCATE — O 8.

Calved May 7, 1881; breeder Mr. A. Taylor; owner Mr. T. H. Harrison; s. King Charles 329; d. Jewel 281 by Rufus 3rd 186; gr d. Mary Grey—O 8.

2283   JUNO — B 9.

Calved 1881; breeder Mr. F. D. Kent; s. Young Prince 608; d. June Rose 968 by The Baron 10; gr d. Rose—B 9.

2284   JUNO — I 2.

Calved December 28, 1881; breeder Mr. T. Brown; s. Priam 373; d. Juliet 967 by Norfolk Duke 127; gr d. Jenny 279 by The Peer 139. See Ruby 4th—I 2 by Tenant Farmer 213.

2285   KATE 2ND — L 12.

Calved January, 1882; breeder Sir E. C. Kerrison, Bart.; s. Lofty 515; d. Kate 971 by Lord Stoke 343; gr d. Kate—L 12.

2286   KNOCK-IN — B 20.

Calved 1880; breeder Mr. A. J. Smith; s. Pickwick 720; d. Loo 2324 by Iron Duke 125; gr d. Piquet 1076 by The Baron 10. See Picket—B 20.

2287   LADY — W 3.

Calved July 21, 1880; breeder Mr. Lofft; owner Rev. H. Evans-Lombe; s. Hector 319; d. Newbourn Pride 5th 1706 by Honest Tom 88; gr d. Newbourn Pride 2nd 384 by Glatton 79. See Newbourn Pride—W 3 by Garibaldi 73.

REGISTER OF COWS.

2288 LADY-BIRD — I 18.
Calved October, 1882; breeder Mr. W. Hudson; s. The Doctor 486; d. Lucy 3rd 1644 by Punch 610; gr d Lucy 2nd 1643 by Davyson 5th 287. See Lucy—I 18 by Proud 547.

2289 LADY-DAY — H 1.
Calved May 13, 1881; breeder the Right Hon. Lord Hastings; s. Davyson 7th 476; d. Davy 17th 846 by Redjacket 7th 169; gr d. Davy 4th 166 by Tenant Farmer 213. See Davy 3rd—H 1 by Sir Nicholas 202.

2290 LADY FANNY CASE — 3 NORF.
Calved January 26, 1882; breeder Mr. John Baly; s. Brutus Duo 463; d. Nancy 2400 by an Elmham Bull.

2291 LADY FULCHER — A 27.
Calved May, 1880; breeder Mr. T. Fulcher; owner Mr. John Baly; s. Brutus 269; d. Brilliant 728 by The Palmer 138; gr d. Brindy 729 by Hero 2nd 86. See Curson—A 27.

2292 LADY HANDSOME — B 8.
Calved June 29, 1882; breeder Her Grace the Duchess of Hamilton; s. The Wilby Lad 599; d. Handsome Lady 241 by Seneca 195; gr d. Handsome—B 8.

2293 LADY JANE — U 45.
Calved November 30, 1882; breeder Mr. C. Austin; s. Rinaldo 556; d. Constance 2nd 800 by Cherry Duke 32; gr d. Constance 799 by Plowman 371. See Weasel—U 45 by Newcastle Prize 359.

2294 LADY KERRISON — O 1.
Calved January 24, 1882; breeder Mr. John Baly; s. The Parson 533; d. Lady Caroline 1610 by Roundhead 400; gr d. Ladybird 977 by Major 109. See Duchess of Suffolk—O 1.

2295 LADY LEGGE — A 11.
Calved July 11, 1882; breeder Mr. John Baly; s. Brutus Duo 463; d. Fortuna 1524 by Redhead 2nd 553; gr d. Frolic 906 by The Palmer 138. See Fanny Bradfield—A 11 by Money 352.

REGISTER OF COWS.

2296 LADY MABEL — I 21.

Calved March, 1882; breeder the Right Hon. the Earl of Leicester, K.G.; owner Mr. W. Hudson; s. The Doctor 486; d. Sepsy 1840 by Proud 547; gr d. Sepsy—I 21 by Redjacket 6th 168.

2297 LADY MACBETH — U 26.

Calved January 7, 1883; breeder Mr. W. Harvey; s. Hamlet 500; d. Dairymaid 2nd 1429 by Bridegroom 266; gr d. Dairymaid—U 26 by Timworth 420.

2298 LADY NELLY — W 3.

Calved May 31, 1882; breeder Mr. John Baly; s. Brutus Duo 463; d. Eleanor 1478 by Davyson 3rd 48; gr d. Helene 945 by Powell 148. See Helen—W 3 by Norfolk Duke 127.

2299 LADY PECK — 4 NORF.

Calved March, 1881; breeder Mr. John Baly; s. Brutus Duo 463; d. Peck by Young Major 235; gr d. by an Elmham Bull. See Introduction.

2300 LADY SONDES — A 27.

Calved September 7, 1882; breeder Mr. John Baly; s. Brutus Duo 463; d. Lady Fulcher 2291 by Brutus 269; gr d. Brilliant 728 by Hero 2nd 86. See Brindy—A 27 by Hero 2nd 86.

2301 LASSIE — L 11.

Calved September 10, 1880; breeder Rev. H. Evans-Lombe; s. Lord Elmham 519; d. Tulip; gr d. Letton Cow—L 11.

2302 LAST BEE — B 12.

Calved October 14, 1882; breeder Mr. H. Biddell; s. Shylock 572; d. Busy Bee 733 by The Baron 10; gr d. Queen Bee 450 by Seneca 195. See The Bee—B 12.

2303 LAURA — A 22.

Calved January 18, 1881; breeder the Right Hon. Lord Suffield, K.C.B.; s. Rupert 567; d. Lilian 1622 by Norfolk 361; gr d. Lily 999 by Witton 432. See Alice—A 22 by Hero of Newcastle 85.

2304   LAURA — A 30.

Calved March 1, 1881; breeder Mr. Garrett Taylor; s. Rufus 188; d. Laurel-leaf 1616 by Osman 530; gr d. Laurel 993 by Morton 353. See Caroline—A 30.

2305   LAURA — P 6.

Calved October 22, 1881; breeder Mr. N. Powell; s. Norfolk John 2nd 527; d. Lively 1633 by Norfolk John 2nd 527; gr d. Lilian 1623 by Norfolk John 131. See Lily—P 6 by Redjacket 7th 169.

2306   LAUREL — I 18.

Calved November, 1881; breeder Mr. W. Hudson; s. The Doctor 486; d. Lucy 2nd 1643 by Davyson 5th 287; gr d. Lucy 1st 1642 by Proud 547. See Lucy—I 18 by Redjacket 6th 168.

2307   LEAF — L 11.

Calved July 12, 1882; breeder the Rev. H. Evans-Lombe; s. Lord Bylaugh 690; d. Nancy; gr d. Letton Cow—L 11.

2308   LEECH — L 11.

Calved July 15, 1882; breeder the Rev. H. Evans-Lombe; s. Lord Bylaugh 690; d. Tulip; gr d. Letton Cow—L 11.

2309   LETTICE — 1 Norf.

Calved October, 1879; breeder Mr. J. Rivett; owner Mr. W. Bradfield; s. Falstaff 303; d. Lucy 2338 by an Elmham Bull; gr d. a Pond Cow.

2310   LIDA — L 11.

Calved December 26, 1882; breeder the Rev. H. Evans-Lombe; s. Lord Bylaugh 690; d. Likely 2312 by Lord Elmham 519; gr d. Mary.

2311   LIGHT HAIR — B 10.

Calved October 13, 1882; breeder Mr. H. Biddell; s. Monarch 4th 351; d. Auburn-locks 1306 by Monarch 4th 351; gr d. Silver-locks 551 by The Baron 10. See Silverbury—B 10 by Playford Sire 142.

2312   LIKELY — L 11.

Calved November 1, 1880; breeder the Rev. H. Evans-Lombe; s. Lord Elmham 519; d. Mary; gr d. Letton Cow—L 11.

2314   LILY — N 2.

Calved October 24, 1882; breeder Her Grace the Duchess of Hamilton; s. The Suffolk Baronet 583; d. Milly 2nd 1670 by Davyson 3rd 48; gr d. Milly 1020 by Powell 143. See Lily 2nd—N 2 by Hero 3rd 87.

2315   LILY OF GLANDFORD — P 6.

Calved February, 1882; breeder Mr. N. Powell; s. Norfolk John 2nd 527; d. Lilian 1623 by Norfolk John 131; gr d. Lily 312 by Redjacket 7th 169. See Primrose—P 6.

2316   LILY — P 6.

Calved October 10, 1881; breeder the Right Hon. Lord Hastings; s. Davyson 6th 475; d. Lily 312 by Redjacket 7th 169; gr d. Primrose 438 by a Red Polled Bull. See Nancy 2nd—P 6 by Norfolk Duke 127.

2317   LINDA — L 11.

Calved December 23, 1882; breeder the Rev. H. Evans-Lombe; s. Lord Bylaugh 690; d. Lassie 2301 by Lord Elmham 519; gr d. Tulip.

2318   LITTLE.

Probationer—Calved December 12, 1880; breeder the Rev. H. Evans-Lombe; s. Lord Elmham 519; d. Swanton.

2319   LITTLE LADY — I 18.

Calved November, 1882; breeder Mr. W. Hudson; s. The Doctor 486; d. Lucy 1st 1642 by Proud 547; gr d. Lucy—I 18 by Redjacket 6th 168.

2320   LITTLE NELL — W 3.

Calved May 15, 1882; breeder Mr. R. E. Lofft; owner the Right Hon. W. H. Smith, M.P.; s. Doubtful 487; d. Newbourn Pride 3rd 1050 by John Bull 820; gr d. Newbourn Pride 383 by Garibaldi 73. See Nelly—W 3 by Robinson 178.

REGISTER OF COWS.

2321 LIVELY 2ND — I 12.

Calved January 23, 1882; breeder Rev. H. Evans-Lombe; s. Lord Elmham 519; d. Lively 1632 by Robin Hood 393; gr d. Caroline 750 by Rufus 189. See Cowslip—I 12.

2322 LOFTY — 1 NORF.

Calved 1878; breeder Mr. W. Wiffin; owner Mr. John Baly; s. an Elmham Bull; d. a Pond Cow. See Introduction.

2323 LONG TAIL — 3 NORF.

Calved 1878; breeder Mr. J. Nicholson; owner Mr. John Baly; s. an Elmham Bull; d. an Elmham Cow. See Introduction.

2324 LOO — B 20.

Calved March 7, 1878; breeder Mr. H. Biddell; owner Mr. A. J. Smith; s. Iron Duke 125; d. Piquet 1076 by The Baron 10; gr d. Picket—B 20.

2325 LOTTIE — L 11.

Calved December 3, 1880; breeder the Rev. H. Evans-Lombe; s. Lord Elmham 519; d. Daisy; gr d. Letton Cow—L 11.

2326 LOTUS — N 7.

Calved July, 1881; breeder Mr. R. H. Mason; s. Slasher 577; d. Lupin 1648 by King Cole 330; gr d. Tulip 1917 by Bradfield 264. See Skelton—N 7 by Necton 3rd 122.

2327 LOUIE — I 12.

Calved December 14, 1881; breeder the Rev. H. Evans-Lombe; s. Lord Elmham 519; d. Lovely 1636 by Robin Hood 393; gr d. Caroline 750 by Rufus 189. See Cowslip—I 12.

2328 LOUISE — L 11.

Calved December, 1882; breeder the Rev. H. Evans-Lombe; s. Lord Bylaugh 690; d. Lottie 2325 by Lord Elmham 519.

REGISTER OF COWS.

2329   LOVELY – A 22.

Calved December 4, 1879; breeder the Right Hon. Lord Suffield, K.C.B.; s. Norfolk 361; d. Lily 999 by Witton 432; gr d. Alice—A 22 by Hero of Newcastle 85.

2330   LOVELY – 2 SUFF.

Calved 1877; breeder Mr. E. Boon; s. a Thornham Bull; d. Nancy by a Kettleburgh (Turner's) Bull; gr d. a Brandon Cow. See Introduction.

2331   LOVELY DUCHESS – V 1.

Calved August 2, 1882; breeder the Most Hon. the Marquis of Bristol; s. Horringer Duke 504; d. Duchess Lovely 1471 by Troston Duke 594; gr d. Lady-love 1613 by Promise 157. See Lovely—V 1 by Doncaster 50.

2332   LOVELY QUEEN – V 1.

Calved January 31, 1882; breeder the Most Hon. the Marquis of Bristol; s. Fancy King 491; d. Lady-love 1613 by Promise 157; gr d. Lovely 324 by Doncaster 50. See Beauty—V 1 by Wonder 230.

2334   LUCKY – L 11.

Calved December 26, 1881; breeder the Rev. H. Evans-Lombe; s. Lord Elmham 519; d. Handsome; gr d. Letton Cow—L 11.

2335   LUCRETIA – AA 1.

Calved September 18, 1881; breeder Mr. G. F. Taber; s. Champion 271; d. Ravinewood Lass 455 by Robin 176; gr d. Nelly 371 by Hero 2nd 86. See Primrose—A 1 by Elmham Sire 67.

2336   LUCRETIA – E 2.

Calved February 27, 1882; breeder Mr. T. Brown; s. Priam 373; d. Lois 1635 by Norfolk Duke 127; gr d. Lady 287 by Old Tom 134. See Red Rose 2nd—E 2 by Stoke 208.

2337   LUCY 2ND – P 6.

Calved October 5, 1882; breeder Mr. N. Powell; s. Premier 543; d. Lively 1633 by Norfolk John 2nd 527; gr d. Lilian 1623 by Norfolk John 131. See Lily—P 6 by Redjacket 7th 169.

2338　　　　　LUCY — 1 Norf.

Calved 1872; breeder Mr. W. Wiffen; owner Mr. J. Rivett; s. an Elmham Bull; d. a Pond Cow. See Introduction.

2339　　　　LUCY GLITTERS — A 20.

Calved February 23, 1882; breeder Mr. J. J. L. Lubbock; s. Quimbo 549; d. Lucy 1010 by Joseph 91; gr d. an Elmham Cow—A 20.

2340　　　　　LUCY JANE — A 20.

Calved January 11, 1883; breeder Mr. J. J. L. Lubbock; s. Roundhead 564; d. Lucy 1010 by Joseph 91; gr d. an Elmham Cow—A 20.

2341　　　　　LUPINE — U 34.

Calved October 6, 1882; breeder Mr. T. L. Palmer; s. Alonso 447; d. Beauty 1313 by Victor 596; gr d. Ruth—U 34 by Timworth 420.

2342　　　　　LYDIA — P 1.

Calved October 15, 1881; breeder Mr. J. J. Colman, M.P.; owner Mr. Garrett Taylor; s. Rufus 188; d. Hetty 1569 by Rufus 188; gr d. Lydia 2nd 1011 by Powell 143. See Handsome 3rd—P 1 by Tenant Farmer 213.

2343　　　　　MABEL — A 2.

Calved January 14, 1881; breeder the Right Hon. Lord Suffield, K.C.B.; s. Rupert 567; d. Judy 966 by Witton 432; gr d. Daffodil 815 by Hero of Newcastle 85. See Cherry—A 2.

2344　　　　　MABEL — P 9.

Calved December 12, 1881; breeder Mr. N. Powell; s. Norfolk John 2nd 527; d. Milkmaid 1668 by Norfolk John 131; gr d. Moss Rose 1683 by Redjacket 7th 169. See Cherry—P 9.

2345　　　　MADAME NICHOLSON.

Probationer—Breeder Mr. Nicholson; owner Mr. Fulcher.

2346　　　　MAGDALEN — 2 Norf.

Calved 1879; breeder Mr. F. J. Mann. See Introduction.

2347   MAGGIE — I 19.

Calved April, 1881; breeder Mr. W. Hudson; s. The Doctor 486; d. Margaret 1654 by Quarles Duke 548; gr d. by Proud 547. See Margaret—I 19.

2348   MAGGIE — P 9.

Calved February 5, 1882; breeder Mr. Nicholas Powell; s. Norfolk John 2nd 527; d. Mildred 1667 by Norfolk John 2nd 527; gr d. Marian 1659 by Norfolk John 181. See Lily—P 9 by Redjacket 7th 169.

2349   MAID MARIAN — 2 NORF.

Calved 1877; breeder Mr. F. J. Mann. See Introduction.

2350   MAID MARIAN 2ND — 2 NORF.

Calved 1881; breeder Mr. F. J. Mann; s. a Red Bull; d. Maid Marian 2349.

2351   MAID MARIAN 3RD — 2 NORF.

Calved April 7, 1883; breeder Mr. F. J. Mann; s. a Red Bull; d. Maid Marian 2349.

2352   MARCHIONESS — L 9.

Calved June 21, 1881; breeder Mr. Jno. Margarson; owner Mr. R. H. Mason; s. Slasher 577; d. Cherry 3rd 1376 by Lord of the Manor 338; gr d. Cherry 770. See Cherry—L 9.

2353   MARGERY — L 9.

Calved June, 1881; breeder Mr. Jno. Margarson; owner Mr. R. H. Mason; s. Purl 611; d. Cheerful 763 by Master Freeman 347; gr d. Cherry 770. See Cherry—L 9.

2354   MARGERY — 2 NORF.

Calved 1876; breeder Mr. F. J. Mann. See Introduction.

2355   MARGUERITE 3RD — N 4.

Calved August 25, 1882; breeder Mr. B. M. Haggard; s. Anthony 448; d. Marguerite 2nd 1658 by Popgun 542; gr d. Marguerite 1657 by King Cole 330. See Daisy 2nd—N 4 by Prince Charlie 151.

REGISTER OF COWS.

2356   MARHAM — M 2.

Calved December 27, 1882; breeder Mr. Garrett Taylor; s. Priam 373; d. Hester 267 by The Peer 139; gr d. Rose 5th 475 by Tenant Farmer 213. See Red Rose—M 2.

2357   MARIA — AA 12.

Calved February 11, 1882; breeder Mr. G. F. Taber; owner Mr. G. K. Taber; s. Champion 271; d. Martha 1662—AA 12 by Ravinewood Beau 160; gr d. Ocean Maid 401 by Hero 3rd 87. See Handsome—A 12.

2358   MARIA — 2 Norf.

Calved 1875; breeder Mr. F. J. Mann. See Introduction.

2359   MARJORAM — D 3.

Calved October 8, 1881; breeder Mr. B. Stimpson; s. Robin Hood 393; d. Miss Marjorie 2nd 1679 by Rob Roy 395; gr d. Miss Marjorie 1025 by Robin 176. See Dame Marjorie—D 3 by an Elmham Bull.

2360   MATTIE — 2 Norf.

Calved 1878; breeder Mr. F. J. Mann. See Introduction.

2361   MEADOW SWEET — B 6.

Calved October 20, 1881; breeder the Most Hon. the Marquis of Bristol; s. Fancy King 491; d. Sweetheart 1883 by Earl of Suffolk 297; gr d. Meadow Sweet 338 by Cremorne 42. See Sweet—B 6 by Seneca 195.

2362   MELTON DAVY 2nd — H 1.

Calved July 10, 1881; breeder the Right Hon. Lord Hastings; s. Thornham Duke 2nd 585; d. Davy 12th 174 by The Baron 9; gr d. Davy 5th 167 by Tenant Farmer 213. See Davy—H 1.

2363   MELTON DAVY 3rd — H 1.

Calved July 23, 1882; breeder the Right Hon. Lord Hastings; s. Roscoe 559; d. Davy 12th 174 by The Baron 9; gr d. Davy 5th 167 by Tenant Farmer 213. See Davy—H 1.

2364 MELTON ROSE – P 7.

Calved September 29, 1880; breeder the Right Hon. Lord Hastings; s. Thornham Duke 2nd 585; d. Rosebud 1804 by Norfolk John 131; gr d. Rose 481 by Redjacket 7th 169. See Polly—P 7.

2365 MELTON ROSE 2ND – P 7.

Calved August 29, 1881; breeder the Right Hon. Lord Hastings; s. Thornham Duke 2nd 585; d. Rosebud 1804 by Norfolk John 131; gr d. Rose 481 by Redjacket 7th 169. See Polly—P 7.

2366 MERCY – A 1.

Calved December 20, 1880; breeder Mr. T. Brown; s. Priam 373; d. Maggie 329 by The Peer 139; gr d. Margaret 331 by Tenant Farmer 213. See Marguerite—A 1 by Hero of Newcastle 85.

2367 MILKMAID – P 4.

Calved 1878; breeder Mr. H. Birkbeck; owner Sir J. W. C. Hartopp, Bart.; s. Roundhead 180; d. Nina 5th 1053 by Norfolk Duke 127; gr d. Nina 3rd 890 by Farmer 70. See Nina 2nd—P 4 by Tenant Farmer 213.

2368 MINNIE 7TH – N 2.

Calved October 20, 1880; breeder Mr. R. E. Lofft; s. Ross 562; d. Minnie 5th 1673 by Bright 267; gr d. Minnie 3rd 343 by Hammond 81. See Minnie—N 2 by Necton Prize 120.

2369 MINNIE 8TH – N 2.

Calved October 27, 1880; breeder Mr. R. E. Lofft; s. Stout 581; d. Minnie 3rd 343 by Hammond 81; gr d. Minnie—N 2 by Necton Prize 120.

2370 MINNIE 9TH – N 2.

Calved December 15, 1881; breeder Mr. R. E. Lofft; s. Long 516; d. Minnie 3rd 343 by Hammond 81; gr d. Minnie—N 2 by Necton Prize 120.

2371 MINNIE – P 9.

Calved February 1, 1882; breeder Mr. N. Powell; s. Norfolk John 2nd 527; d. Molly 1682 by Norfolk John 131; gr d. Moss Rose 1683 by Redjacket 7th 169. See Cherry—P 9.

2372   MIRANDA — AK 17.

Calved September 2, 1882; breeder Mr. G. F. Taber; s. Red Knight 735; d. Princess May 1754 by Crown Prince 281; gr d. Thornham Princess 1230 by Eclipse 2nd 299. See Thursford Queen—K 17 by Tenant Farmer 213.

2373   MIRIAM — W 3.

Calved October 13, 1881; breeder Mr. R. E. Lofft; owner Mr. C. Austin; s. Long 516; d. Newbourn Pride 7th 1708 by Donald 291; gr d. Newbourn Pride 3rd 1050 by John Bull 326. See Newbourn Pride—W 3 by Garibaldi 73.

2374   MIRTH 3RD — O 6.

Calved February, 1881; breeder Sir E. C. Kerrison, Bart.; s. Lofty 515; d. Mirth 2nd 1678 by King Charles 329; gr d. Mirth 344 by Rifleman 175. See Vanity—O 6.

2375   MIRTH 4TH — O 6.

Calved November, 1882; breeder Sir E. C. Kerrison, Bart.; s. Lofty 515; d. Mirth 2nd 1678 by King Charles 329; gr d. Mirth 344 by Rifleman 175. See Vanity—O 6.

2376   MISSIE — 2 NORF.

Calved October, 1882; breeder Mr. F. J. Mann; d. Maria 2358.

2377   MISS BAKER — 1 SUFF.

Calved January 20, 1882; breeder Mr. G. Gooderham; s. Wild Robin 600; d. Baker 1992.

2378   MISS BELL — X 5.

Calved May 21, 1882; breeder Her Grace the Duchess of Hamilton; s. The Suffolk Baronet 583; d. Blue Bell—X 5 by Youngster 439.

2379   MISS EMMA — A 4.

Calved February 2, 1883; breeder Mr. John Baker; s. Playmate 723; d. Cassie 1363 by Priam 373; gr d. Ringlet 2nd 465 by Tenant Farmer 213. See Ringlet—A 4 by Hero of Newcastle 85.

REGISTER OF COWS.

2380   MISS MARIA — 2 Norf.
Calved 1879; breeder Mr. F. J. Mann; d. Maria 2358.

2381   MISS MATTIE — 2 Norf.
Calved June 25, 1882; breeder Mr. F. J. Mann; d. Mattie 2360.

2382   MISTLETOE — A 1.
Calved October 30, 1882; breeder Mr. T. Brown; s. Priam 373; d. Maggie 329 by The Peer 139; gr d. Margaret 831 by Tenant Farmer 213. See Marguerite—A 1 by Hero of Newcastle 85.

2383   MOSS ROSE — B 9.
Calved November 22, 1882; breeder Mr. H. Biddell; s. Shylock 572; d. Standard Rose 1867 by Crown Prince 281; gr d. Rose—B 9.

2384   MRS. CASE — 1 Norf.
Calved 1879; breeder Mr. W. Wiffin; owner Mr. John Baly; s. an Elmham Bull; d. a Pond Cow. See Introduction.

2385   MRS. NICHOLSON — 3 Norf.
Calved 1879; breeder Mr. J. Nicholson; owner Mr. John Baly; s. an Elmham Bull; d. an Elmham Cow. See Introduction.

2386   MY LADY — P 3.
Calved January 28, 1882; breeder Mr. Garrett Taylor; s. Cato 468; d. Duchess 1468 by Duke of Norfolk 295; gr d. Rose 2nd 479 by Tenant Farmer 213. See Rose—P 3.

2387   MYRTLE — A 1.
Calved December 1, 1882; breeder Mr. T. Brown; s. Priam 373; d. Marian 1012 by Arthur 4; gr d. Maggie 329 by The Peer 139. See Margaret—A 1 by Tenant Farmer 213.

2388   MYSTERIOUS — R 1.
Calved January 18, 1881; breeder Mr. Garrett Taylor; s. Grey Spot 498; d. Sophia 2543 by Trimmer 218; gr d. Sweetmeat 594 by Young Duke 234. See Susan—R 1 by Tommy 216.

REGISTER OF COWS.

2389 MYSTERY — R 1.

Calved December 28, 1879; breeder Mr. Garrett Taylor; s. Grey Spot 498; d. Sophia 2543 by Trimmer 218; gr d. Sweetmeat 594 by Young Duke 234. See Susan—R 1 by Tommy 216.

2390 MYSTIC — R 1.

Calved February 28, 1882; breeder Mr. Garrett Taylor; s. Davyson 8th 477; d. Sophia 2543 by Trimmer 218; gr d. Sweetmeat 594 by Young Duke 234. See Susan—R 1 by Tommy 216.

2391 NAN OF ELMHAM — A 12.

Calved September 1882; breeder Mr. Jno. Howling; s. Tommy 588; d. Nancy 2392 by Brutus 269; gr d. Boulter 54 by Hero 3rd 87. See Handsome—A 12.

2392 NANCY — A 12.

Calved 1877; breeder Mr. Jno. Howling; s. Brutus 269; d. Boulter 54 by Hero 3rd 87; gr d. Handsome—A 12.

2393 NANCY 3RD — P 9.

Calved February 8, 1882; breeder Mr. N. Powell; s. Norfolk John 2nd 527; d. Marian 1659 by Norfolk John 131; gr d. Lucy 325 by Redjacket 7th 169. See Cherry—P 9.

2394 NANCY — R 9.

Calved November, 1874; breeder Mr. T. Easter; owner Mr. J. Baker; s. Read 385; d. Brundish—R 9 by a Laxfield Bull.

2395 NANCY — R 11.

Calved January 2, 1882; breeder Mr. W. B. Easter; s. Brundish Prince 462; d. Pretty 1092 by Read 385; gr d. Pretty—R 11 by a Laxfield Bull.

2396 NANCY — M 2.

Calved July 24, 1881; breeder Mr. R. E. Lofft; owner Mr. C. Austin; s. Doubtful 487; d. Helena 944 by Benedict 17; gr d. Rose 4th 474 by Tenant Farmer 213. See Red Rose—M 2.

2397   NANCY 7TH — T 18.

Calved January 9, 1881; breeder Mr. Jno. Howell; s. Davyson 9th 478; d. Nancy 1039 by Masker 346; gr d. The Nun—T 18.

2398   NANCY 8TH — T 18.

Calved April 12, 1882; breeder Mr. Jno. Howell; s. Davyson 9th 478; d. Nancy 2nd 1693 by Sir Thomas 489; gr d. Nancy 1039 by Masker 346. See The Nun—T 18.

2399   NANCY — 1 NORF.

Calved 1877; breeder Mr. Wiffen; owner Mr. J. Rivett; s. an Elmham Bull; d. a Pond Cow. See Introduction.

2400   NANCY — 3 NORF.

Calved 1876; breeder Mr. J. Nicholson; owner Mr. Jno. Baly; s. an Elmham Bull; d. an Elmham Cow. See Introduction.

2401   NANCY 2ND — 2 SUFF.

Calved 1879; breeder Mr. E. Boon; s. a Wolton Bull; d. Cherry 2nd 2181 by a Thornham Bull; gr d. Cossett by an Oakley Bull.

2402   NANNY — K 19.

Calved August 9, 1881; breeder Mr. W. A. T. Amherst, M.P.; s. Davyson 3rd 48; d. Nancy 2nd 1691 by Young Major 235; gr d. Nancy 1690 by Peck 534. See Spot 3rd—K 19 by Wilby Chapman 228.

2403   NANZIE — O 6.

Calved April 12, 1881; breeder Sir J. W. C. Hartopp, Bart.; s. Hardwick 501; d. Canteen 1361 by a Son of Rifleman 175; gr d. Mirth 344 by Rifleman 175. See Vanity—O 6.

2404   NECTARINE — N 6.

Calved June 20, 1881; breeder the Right Hon. Lord Hastings; s. Thornham Duke 2nd 585; d. Dainty 819 by Prince Charlie 151; gr d. Nancy 359 by Fransham Captain 71. See Tit—N 6 by Necton 3rd 122.

REGISTER OF COWS.

2405 NECTARINE 2ND — N 6.
Calved May 27, 1882; breeder the Right Hon. Lord Hastings; s. Davyson 7th 476; d. Dainty 819 by Prince Charlie 151; gr d. Nancy 359 by Fransham Captain 71. See Tit—N 6 by Necton 3rd 122.

2406 NELLY — 1 NORF.
Calved September, 1881; breeder Mr. J. Rivett; s. Falstaff 303; d. Nancy 2399 by an Elmham Bull; gr d. a Pond Cow.

2407 NELLY — 2 SUFF.
Calved July, 1882; breeder Mr. E. Boon; s. Wild Roger 603; d. Nancy 2nd 2401 by a Wolton Bull; gr d. Cherry 2nd 2081 by a Thornham Bull.

2408 NEST — 1 NORF.
Calved August, 1880; breeder Mr. J. Rivett; s. Falstaff 303; d. Nancy 2399 by an Elmham Bull; gr d. a Pond Cow.

2409 NEWBOURN PRIDE 11TH — W 3.
Calved December 5, 1881; breeder Mr. R. E. Lofft; s. Rollick 558; d. Newbourn Pride 9th 1710 by Stout 581; gr d. Newbourn Pride 5th 1706 by Honest Tom 88. See Newbourn Pride 2nd—W 3 by Glatton 79.

2410 NIGHTINGALE — A 6.
Calved November 25, 1882: breeder Mr. H. le Strange; s. Goshawk 497; d. Redbreast 1766 by an Elmham Bull; gr d. Norton 392 by Hero 3rd 87. See Norton—A 6.

2411 NINA — A 24.
Calved May 14, 1881; breeder the Right Hon. Lord Suffield, K.C.B.; owner Mrs. Collyer; s. Rupert 567; d. Flora 1519 by Norfolk 366; gr d. Floss 899 by Witton 432. See Floss—A 24.

2412 NINA — A 33.
Calved November 8, 1882; breeder Mr. R. H. Mason; s. Slasher 577; d. Nancy 1688 by King Tom 335; gr d. Elm-leaf 2nd 1489 by King Tom 335. See Elm-leaf—A 33.

2413 NIPPENOSE BELLE — AA 12.

Calved March 4, 1882; breeder Mr. G. F. Taber; owner Mr. G. L. Sanderson; s. Champion 271; d. Mollie 1681—AA 12 by Ravinewood Beau 160; gr d. Ocean Maid 401 by Hero 3rd 87. See Handsome—A 12.

2414 NONSENSE — S 4.

Calved January 15, 1882; breeder Mr. A. Taylor; s. Starston Duke 570; d. Novel 1056 by Easton Duke 61; gr d. Novelty 395 by Tommy 216. See Holkham—S 4 by Sporle.

2415 NORA — A 1.

Calved December 11, 1880; breeder the Right Hon. Lord Suffield, K.C.B.; owner Mr. C. Waters; s. Norfolk 361; d. Polly 1081 by Witton 432; gr d. Polly by Redjacket 2nd 164. See Primrose—A 1 by Elmham Sire 67.

2416 NORA — AN 7.

Calved March 24, 1882; breeder Mr. G. F. Taber; owner Mr. J. L. Mustard; s. Champion 271; d. Susie 1220—AN 7 by Ravinewood Beau 160; gr d. Skelton—N 7 by Necton 3rd 122.

2417 NOSEGAY — A 22.

Calved April 16, 1881; breeder the Right Hon. Lord Suffield, K.C.B.; owner Rev. J. C. Girling; s. Rupert 567; d. Alice 671 by Witton 432; gr d. Alice—A 22 by Hero of Newcastle 85.

2418 NOSEGAY 4TH — O 5.

Calved January, 1881; breeder Sir E. C. Kerrison, Bart.; s. Lofty 515; d. Nosegay 3rd 1055 by Harold 83; gr d. Nosegay 393 by Rifleman 175. See Nosegay—O 5.

2419 NOSEGAY 5TH — O 5.

Calved November, 1882; breeder Sir E. C. Kerrison, Bart.; s. Lofty 515; d. Nosegay 3rd 1055 by Harold 83; gr d. Nosegay 393 by Rifleman 175. See Nosegay—O 5.

2420 NOVICE — R 2.

Calved May 18, 1882; breeder Mr. A. Taylor; s. Starston Duke 570; d. Nancy 363 by Richard II. 173; gr d. Lovely 2nd 822 by Richard II. 173. See Pretty—R 2 by Richard I. 172.

REGISTER OF COWS.

2421 THE NUN – W 9.

Calved May 22, 1882; breeder Mr. H. Biddell; s. The Friar 495; d. Grand Lady 1547 by Monarch 4th 351; gr d. Little Lady 1004 by The Baron 10. See Lady—W 9.

2422 NUTMEG – 1 NORF.

Calved November 22, 1882; breeder Mr. J. Rivett; s. Falstaff 303; d. Nancy 2399 by an Elmham Bull; gr d. a Pond Cow.

2423 OAKLEY 4TH – O 1.

Calved 1879; breeder Sir E. C. Kerrison, Bart.; s. Roundhead 400; d. Oakley 3rd 400 by Major 109; gr d. Oakley 398 by Rifleman 175. See Duchess of Suffolk—O 1.

2424 OLGA – U 23.

Calved 1875; breeder Mr. W. Harvey; s. a Timworth Red Bull; d. a Timworth Red Cow. See Annie—U 23.

2425 OLLY – H 4.

Calved January 16, 1882; breeder Mr. Garrett Taylor; s. Grey Spot 498; d. Olive 2nd 1714 by Romulus 398; gr d. Olive—H 4.

2426 OPHELIA – U 31.

Calved January 8, 1883; breeder Mr. W. Harvey; s. Hamlet 500; d. Rosamond 1139 by Timworth 420. See Rosamond—U 31.

2427 PASTILLE – P 1.

Calved March 19, 1882: breeder Mr. T. Brown; s. Bergamot 455; d. Penelope 1069 by Roundhead 180; gr d. Nelly 372 by Redjacket 7th 169. See Handsome 2nd—P 1 by Tenant Farmer 213.

2428 PATCHWORK – E 11.

Calved March, 1881; breeder Mr. T. Fulcher; s. Redhead 2nd 553; d. Pansy 1063 by Cringleford Duke 43; gr d. Pretty 422 by Cantley 29. See Polly—E 11.

2429   PATTIE — B 20.

Calved January 4, 1882; breeder Mr. W. A. T. Amherst, M.P.; s. St. Edmund 580; d. Pansie 1717 by Crown Prince 281; gr d. Picotee 410 by Great Britain 80. See Picket—B 20.

2430   PATTIE 2ND — B 20.

Calved January 4, 1883; breeder Mr. W. A. T. Amherst, M.P.; s. Davyson 3rd 48; d. Pansie 1717 by Crown Prince 281; gr d. Picotee 410 by Great Britain 80. See Picket—B 20.

2431   PATTY — A 23.

Calved May 4, 1881; breeder the Right Hon. Lord Suffield, K.C.B.; owner Rev. J. C. Girling; s. Rupert 567; d. Dido 850 by Rufus 187; gr d. Winifred 1278 by Witton 432. See Dido—A 23 by Hero of Newcastle 85.

2432   PEACH BLOSSOM — B 13.

Calved June 18, 1877; breeder Mr. H. Biddell; owner Mr. A. J. Smith; s. Iron Duke 125; d. Blossom—B 13 by Powell's of Kelvedon.

2433   PEACH BUD — B 13.

Calved 1880; breeder Mr. A. J. Smith; s. Pickwick 720; d. Peach Blossom 2432 by Iron Duke 125; gr d. Blossom—B 13 by Powell's of Kelvedon.

2434   PEACH LEAF — B 13.

Calved 1881; breeder Mr. A. J. Smith; s. Pickwick 720; d. Peach Blossom 2432 by Iron Duke 125; gr d. Blossom—B 13 by Powell's of Kelvedon.

2435   PEARL I — O 13.

Calved June 9, 1882; breeder the Right Hon. Lord Henniker; s. Cyprus 473; d. Pearl 1727 by Thornham Duke 418; gr d. Ruby 1165 by Ruddy 402. See Strawberry—O 13.

2436   PENAL — V 11.

Calved February, 1882; breeder Mr. J. M. Spinks; s. Lord George 520; d. Penance 1728 by Damian 282; gr d. Penguin 1070 by Monarch 2nd 242. See Gloss 2nd—V 11 by Boss 237.

REGISTER OF COWS.

2437 PENELOPE OF ELMHAM — E 11.

Calved February 13, 1883; breeder Mr. T. Fulcher; s. Tommy 588; d. Pansy 1063 by Cringleford Duke 43; gr d. Pretty 422 by Cantley 29. See Polly—E 11.

2438 PEONY — K 25.

Calved 1882; breeder Mr. P. Blofield; s. Rollick 558; d. Prudence 1115 by Young Major 235; gr d. Princess by a Stoke Bull. See Bride—K 25.

2439 PHILIPPA — N 4.

Calved May, 1882; breeder Mr. R. H. Mason; s. Philip 538; d. Strawberry 2nd 2552 by King Harry 332; gr d. Strawberry 1872 by King Tom 335. See Daisy—N 4 by Necton 3rd 122.

2440 PHŒBE — A 24.

Calved June 18, 1880; breeder the Right Hon. Lord Suffield, K.C.B.; s. Norfolk 361; d. Flora 1519 by Norfolk 361; gr d. Floss 899 by Witton 432. See Floss—A 24.

2441 PHŒBE — K 25.

Calved 1882; breeder Mr. P. Blofield; s. Rollick 558; d. Primrose 1748 by Davyson 3rd 48; gr d. Prune 1116 by Young Major 235. See Princess—K 25 by a Stoke Bull.

2442 PHŒNIX 2ND — U 6.

Calved August 19, 1878; breeder Mr. R. E. Lofft; s. Hector 319; d. Phœnix—U 6 by a Troston Bull.

2443 PICKET — A 1.

Calved January 20, 1881; breeder the Right Hon. Lord Suffield, K.C.B.; owner Mr. C. Waters; s. Rupert 567; d. Primrose 1747 by Norfolk 361; gr d. Polly 1081 by Witton 432. See Polly—A 1 by Redjacket 2nd 164.

2444 PIMPERNEL — A 3.

Calved October 18, 1882; breeder Mr. T. Leonard Palmer; s. Alonso 447; d. Elmham 3rd 1485 by Hector 319; gr d. Elmham 199 by Hero 3rd 87. See Brettenham Handsome—A 3 by Hero of Newcastle 85.

2445   PINE — A 1.

Calved February 23, 1883; breeder Mr. C. Waters; s. Red Knight 735; d. Picket 2443 by Rupert 567; gr.d. Primrose 1747 by Norfolk 361. See Polly—A 1 by Witton 432.

2446   PINK — O 1.

Calved 1880; breeder Sir J. W. C. Hartopp, Bart.; s. Bounty 460; d. Princess Royal 1755 by a Son of Rifleman 175; gr d. Victoria 625 by Rifleman 175. See Duchess of Suffolk—O 1.

2447   POLLY — A 25.

Calved November 5, 1879; breeder the Right Hon. Lord Suffield, K.C.B.; s. Norfolk 361; d. Poppy 1088 by Rufus 187; gr d. Pansy—A 25 by Hero of Newcastle 85.

2448   POLLY — P 7.

Calved October 13, 1882; breeder Mr. N. Powell; s. Premier 543; d. Primrose 3rd 1749 by Norfolk John 131; gr d. Polly 416 by a Red Polled Bull. See Violet—P 7.

2449   POLLY — R 9.

Calved December 15, 1882; breeder Mr. T. R. West; s. Brundish Prince 462; d. Brundish Cow—R 9 by Laxfield Sire.

2450   POLLY — 1 Norf.

Calved 1876; breeder Mr. Wiffen; owner Mr. J. Rivett; s. an Elmham Bull; d. a Pond Cow. See Introduction.

2451   POND LILY — 1 Norf.

Calved 1880; breeder Mr. J. Rivett; owner Mr. T. Fulcher; s. Falstaff 303; d. a Pond Cow. See Introduction.

2452   POND LILY 2ND — 1 Norf.

Calved 1880; breeder Mr. J. Rivett; owner Mr. T. Fulcher; s. Falstaff 303; d. a Pond Cow. See Introduction.

2453   POND LILY 3RD — 1 Norf.

Calved 1880; breeder Mr. J. Rivett; owner Mr. T. Fulcher; s. Falstaff 303; d. a Pond Cow. See Introduction.

2454 POPPET 5TH — U 43.

Calved August 15, 1881; breeder Mr. R. E. Lofft; s. Ross 562; d. Poppet 3rd 1742 by Honest Tom 88; gr d. Poppet—U 43 by Sampson 191.

2455 POPPINETTE — U 43.

Calved January 3, 1883; breeder Mr. W. A. T. Amherst, M.P.; s. Davyson 3rd 48; d. Poppy 2456 by Stout 581; gr d. Poppet 2nd 1087 by Cherry Duke 32. See Poppet—U 43 by Sampson 191.

2456 POPPY — U 43.

Calved July, 1880; breeder Mr. R. E. Lofft; owner Mr. W. A. T. Amherst, M.P.; s. Stout 581; d. Poppet 2nd 1087 by Cherry Duke 32; gr d. Poppet—U 43 by Sampson 191.

2457 POPPY — 1 NORF.

Calved July, 1880; breeder Mr. J. Rivett; s. Falstaff 303; d. Polly 2450 by an Elmham Bull; gr d. a Pond Cow.

2458 PORTIA — B 17.

Calved May 9, 1882; breeder Mr. H. Biddell; s. Shylock 572; d. Pallas 1716 by Iron Duke 125; gr d. Wisdom 656 by Seneca 195. See Fairy—B 17.

2459 PORTIA — U 32.

Calved December 31, 1882; breeder Mr. W. Harvey; s. Hamlet 500; d. Rosebud 2nd 1806 by Victor 596; gr d. Rosebud 1166 by Timworth 420. See Rosebud—U 32.

2460 POWLY — K 17.

Calved May 4, 1881; breeder Sir J. W. C. Hartopp, Bart.; s. Bounty 460; d. Bright Cherry 724 by Peter Powell 370; gr d. Cherry 4th 768 by Norfolk Duke 127. See Cherry 2nd—K 17 by Norfolk Duke 127.

2461 PRESS — V 11.

Calved November, 1880; breeder Mr. J. M. Spinks; owner Mr. Garrett Taylor; s. Lord George 520; d. Penguin 1070 by Monarch 2nd 242; gr d. Gloss 2nd 665 by Boss 237. See Gloss—V 11.

2462   PRETTY – A 25.

Calved January 28, 1881; breeder the Right Hon. Lord Suffield, K.C.B.; owner Mr. C. Waters; s. Rupert 567; d. Pansy 1718 by Norfolk 361; gr d. Poppy 1088 by Rufus 187. See Pansy—A 25 by Hero of Newcastle 85.

2463   PRETTY – R 9.

Calved December 1879; breeder Mr. T. Easter; owner Mr. B. M. Haggard; s. Charles 639; d. Nancy 2394 by Read 385; gr d. Brundish—R 9 by a Laxfield Bull.

2464   PRETTY BIRD – B 4.

Calved June 28, 1882; breeder Mr. H. Biddell; s. The Friar 494; d. Little Bird 1629 by Grand Turk 316; gr d. Barley Bird 686 by Iron Duke 125. See Little Wryneck—B 4 by Playford Sire 142.

2465   PRETTY BLOSSOM – B 18.

Calved September 18, 1882; breeder Mr. W. A. T. Amherst, M.P.; s. The Friar 494; d. Pretty Flower 1093 by Iron Duke 125; gr d. Fancy Flower 219 by Seneca 195. See Fancy—B 18.

2466   PRIDE – V 11.

Calved December, 1881; breeder Mr. J. M. Spinks; s. Lord George 520; d. Princess 1109 by Max 112; gr d. Penguin 1070 by Monarch 2nd 242. See Gloss 2nd—V 11 by Boss 237.

2467   PRIMROSE NELL – A 1.

Calved July, 1881; breeder Mr. W. Bradfield; owner Mr. G. F. Taber; s. Tommy 588; d. Nellie 1702 by The Palmer 138; gr d. Nelly 871 by Hero 2nd 86. See Primrose—A 1 by Elmham Sire 67.

2468   PRIMULA – U 5.

Calved December, 1880; breeder Mr. R. E. Lofft; owner Mr. W. A. T. Amherst, M.P.; s. Stout 581; d. Cauliflower 6th 1366 by Bright 267; gr d. Cauliflower 4th 755 by Cherry Duke 32. See Cauliflower—U 5 by Sampson 191.

REGISTER OF COWS.

2469 PRINCESS C – K 17.
Calved December 30, 1881; breeder the Right Hon. Lord Henniker; s. Cyprus 473; d. Thornham Princess 1230 by Eclipse 2nd 299; gr d. Thursford Queen 599 by Tenant Farmer 213. See Cherry—K 17.

2470 PRINCESS LOVELY – V 1.
Calved March 26, 1882; breeder the Most Hon. the Marquis of Bristol; s. Fancy King 491; d. Lovely 324 by Doncaster 50; gr d. Beauty 36 by Wonder 230. See Cowslip—V 1.

2471 PRINCESS MARGARET – U 29.
Calved July 22, 1882; breeder Mr. W. Harvey; s. Victor 596; d. Red Rose 1126 by Timworth 420. See Red Rose—U 29.

2472 PRISCILLA OF ELMHAM – E 11.
Calved March 17, 1882; breeder Mr. T. Fulcher; s. Lofty 515; d. Pansy 1063 by Cringleford Duke 43; gr d. Pretty 422 by Cantley 29. See Polly—E 11 by Duke 52.

2473 PRUDE – 1 Norf.
Calved October, 1881; breeder Mr. J. Rivett; s. Falstaff 303; d. Polly 2450 by an Elmham Bull; gr d. a Pond Cow.

2474 PRUDISH – T 1.
Calved February 6, 1882; breeder Mr. Garrett Taylor; owner Mr. J. J. Colman, M.P.; s. Cato 468; d. Prune 1757 by Rufus 188; gr d. Primrose 2nd 440 by Farmer 70. See Primrose—T 1 by Tenant Farmer 213.

2475 QUALITY – R 8.
Calved February, 1881; breeder Mr. T. Easter; owner Mr. G. Holmes; s. Simon 408; d. Dingoo 2151 by Read 385; gr d. Beauty—R 8 by a Laxfield Bull.

2476 QUEEN D – K 17.
Calved July 24, 1881; breeder the Right Hon. Lord Henniker; s. Cyprus 473; d. Fairy Queen 1503 by Thornham Duke 418; gr d. Thursford Queen 599 by Tenant Farmer 213. See Cherry—K 17.

2477   QUEEN G — K 17.

Calved May 26, 1882; breeder the Right Hon. Lord Henniker; s. Cyprus 473; d. Fairy Queen 1503 by Thornham Duke 418; gr d. Thursford Queen 599 by Tenant Farmer 213. See Cherry—K 17.

2478   QUEEN — U 23.

Calved 1881; breeder Mr. W. Harvey; s. Victor 596; d. Annie 682 by Timworth 420. See Annie—U 23.

2479   QUEEN MAY — W 3.

Calved July 21, 1881; breeder Mr. R. E. Lofft; owner Mr. C. Austin; s. Doubtful 487; d. Newbourn Pride 4th 1051 by Cherry Duke 32; gr d. Newbourn Pride 2nd 384 by Glatton 79. See Newbourn Pride—W 3 by Garibaldi 73.

2480   RACINE — AA 29.

Calved July 14, 1881; breeder Mr. G. F. Taber; s. Champion 271; d. Rachel 1121—AA 29 by Ravinewood Beau 160; gr d. Ravinewood Belle 454 by Hero 3rd 87. See Mrs. Rollin—A 29.

2481   RAISIN — E 2.

Calved December 26, 1882; breeder Mr. T. Brown; s. Priam 373; d. Rosamond 1136 by Norfolk Duke 127; gr d. Rosebud 489 by Stoke Duke 209. See Rose of Eaton—E 2 by Cringleford Sire 67.

2482   RASPBERRY — N 4.

Calved July, 1881; breeder Mr. R. H. Mason; s. Slasher 577; d. Strawberry 2nd 2552 by King Harry 332; d. Strawberry 1872 by King Tom 335. See Daisy—N 4 by Necton 3rd 122.

2483   RED BEAUTY — V 2.

Calved April 12, 1881; breeder Mr. G. Gooderham; owner General L. F. Ross; s. Wild Rocket 601; d. Red Stockings 2nd 1128 by Councillor 38; gr d. Flora 2nd 897 by Doncaster 50. See Flora—V 2 by King Alfred 96.

2484   RED BEAUTY 2ND — V 2.

Calved April 12, 1881; breeder Mr. G. Gooderham; owner Mr. G. F. Taber; s. Wild Rocket 601; d. Red Stockings 2nd 1128 by Councillor 38; gr d. Flora 2nd 897 by Doncaster 50. See Flora—V 2 by King Alfred 96.

REGISTER OF COWS.

2485   RED BLOSSOM — B 13.
Calved 1880; breeder Mr. A. J. Smith; s. Pickwick 720; d. Black Blossom 2020 by Iron Duke 125; gr d. Blossom—B 13 by Powell's of Kelvedon.

2486   REDCREST — I 4.
Calved October 24, 1882; breeder Mr. H. le Strange; s. Goshawk 497; d. Linnett 1002 by Edgar 64; gr d. Handsome—I 4.

2487   RED DAISY — H 2.
Calved April 1, 1882; breeder Mr. Garrett Taylor; s. Cato 468; d. Easton Daisy 1474 by Skobeloff 573; gr d. Daisy 3rd 823 by Powell 143. See Daisy 1st—H 2 by Young Duke 234.

2488   RED SPOT — A 1.
Calved February 11, 1882; breeder Mr. Garrett Taylor; s. Davyson 8th 477; d. White Spot 1934 by Lord John 340; gr d. Moss Rose 1031 by Powell 143. See Rosebud 2nd—A 1 by Hero 3rd 87.

2489   RED TULIP 2ND — N 5.
Calved December 29, 1881; breeder Mrs. E. Perkins; s. Othello 532; d. Red Tulip 1779 by King Cole 330; gr d. Rose 477 by Prince Charlie 151. See Tulip 2nd—N 5 by Necton 3rd 122.

2490   RHODA — K 26.
Calved January, 1883; breeder Mr. P. Blofield; s. Rollick 558; d. Rosette 1162 by Young Major 235; gr d. Rose by a Stoke Bull. See Fuller—K 26.

2491   RISBY TOPKNOT — W 10.
Calved August 25, 1882; breeder Mr. W. G. Collins; owner Mr. G. J. Paine; s. High Sheriff 676; d. Hunston Topknot 1579 by Oliver 364; gr d. Topknot—W 10 by Duke of Suffolk 57.

2492   ROSA — R 9.
Calved September, 1878; breeder Mr. T. R. West; s. Harold 83; d. Brundish Cow—R 9.

REGISTER OF COWS.

2493   ROSA 2ND — R 9.

Calved July 13, 1882; breeder Mr. T. R. West; s. Brundish Prince 462; d. Rosa 2492 by Harold 83; gr d. Brundish Cow—R 9.

2494   ROSABELLE — B 9.

Calved August 6, 1881; breeder the Most Hon. the Marquis of Bristol; s. Fancy King 491; d. Rosa 1785 by Fancy Duke 490; gr d. Rosary 1140 by Crown Prince 281. See Rosette—B 9 by Cremorne 42.

2495   ROSALIE — K 17.

Calved August 25, 1882; breeder Mr. J. J. Colman, M.P.; s. King Charles 329; d. Rosebud 2nd 1797 by Rufus 188; gr d. Rosebud 494 by Norfolk Duke 127. See Cherry 2nd—K 17 by Tenant Farmer 213.

2496   ROSE — N 6.

Calved December 20, 1882; breeder Mr. H. Haylock; s. The Parson 533; d. Edith 2158 by Rufus 188; gr d. Cherry 94 by Fransham Captain 71. See Tit—N 6 by Necton 3rd 122.

2497   ROSE E — O 14.

Calved January 18, 1882; breeder the Right Hon. Lord Henniker; s. Cyprus 473; d. Rose Crown 1807 by Crown Prince 281; gr d. Rosebud 1154 by Eclipse 68. See Roseleaf—O 14 by Ruddy 402.

2498   ROSE F — O 14.

Calved November 28, 1882; breeder the Right Hon. Lord Henniker; s. Thornham Hero 763; d. Rose Crown 1807 by Crown Prince 281; gr d. Rosebud 1154 by Eclipse 68. See Roseleaf—O 14 by Ruddy 402.

2499   ROSE — S 2.

Calved March 24, 1882; breeder Sir J. W. C. Hartopp, Bart.; s. Hardwick 501; d. Flora 2194 by Bounty 460; gr d. Red Berry 1765 by Roundhead 180. See Harebell—S 2 by Powell 143.

2500 ROSE 2ND OF TRIMLEY — X 1.

Calved January 25, 1882; breeder Mr. C. K. Cordy; s. Ready 551; d. Red Rose 462 by Prince 145; gr d. Rose 483 by King Alfred 96. See Lovely—X 1.

REGISTER OF COWS.

2501   ROSEBUD — B 9.

Calved October 30, 1881; breeder the Most Hon. the Marquis of Bristol; s. Fancy King 491; d. Rosary 1140 by Crown Prince 281; gr d. Rosette 508 by Cremorne 42. See New Rose—B 9 by Seneca 195.

2502   ROSETTE 3RD — N 4.

Calved March 12, 1881; breeder Mrs. E. Perkins; s. Osman 531; d. Rosette 1819 by King Tom 335; gr d. Rose 1142 by Bradfield 264. See Daisy—N 4 by Necton 3rd 122.

2503   ROSETTE 4TH — N 4.

Calved April 14, 1881; breeder Mrs. E. Perkins; s. Osman 531; d. Rosette 2nd 1820 by King Cole 330; gr d. Rose 1142 by Bradfield 264. See Daisy—N 4 by Necton 3rd 122.

2504   ROSETTE 5TH — N 4.

Calved January 22, 1882; breeder Mrs. E. Perkins; s. Othello 532; d. Rosette 1819 by King Tom 335; gr d. Rose 1142 by Bradfield 264. See Daisy—N 4 by Necton 3rd 122.

2505   ROSELEAF 2ND — K 19.

Calved January 15, 1883; breeder Mr. W. A. T. Amherst, M.P.; s. Davyson 3rd 48; d. Rose 484 by Young Major 235; gr d. Spot 2nd by Wilby Chapman 228; 2nd gr d. Spot 558 by Wonder 231. See Rose—K 19 by an Elmham Bull.

2506   ROSELEAF B — O 14.

Calved December 27, 1881; breeder the Right Hon. Lord Henniker; s. Cyprus 473; d. Roseleaf 1159 by Ruddy 402; gr d. Cherry—O 14.

2507   ROSELEAF H — O 14.

Calved May 21, 1882; breeder the Right Hon. Lord Henniker; s. Cyprus 473; d. Roseleaf 2nd 1816 by Thornham Duke 418; gr d. Roseleaf 1159 by Ruddy 402. See Cherry—O 14.

2508 ROSEMARY — P 3.

Calved September 14, 1882; breeder Mr. J. J. Colman, M.P.; s. King Charles 329; d. Rosamond 1789 by Rufus 188; gr d. Rosa 1133 by Norfolk Duke 127. See Rose 3rd—P 3 by Young Duke 234.

2509 ROSE PRINCESS — K 26.

Calved 1879; breeder Mr. P. Blofield; s. Davyson 3rd 48; d. Rosette 1162 by Young Major 235; gr d. Rose by a Stoke Bull. See Fuller—K 26.

2510 ROSINA — N 4.

Calved July, 1881; breeder Mr. R. H. Mason; s. Slasher 577; d. Rosalind 1787 by King Harry 332; gr d. Rose 2nd 1143 by Longham 104. See Daisy—N 4 by Necton 3rd 122.

2511 ROSINETTE — N 4.

Calved April, 1881; breeder Mr. R. H. Mason; s. Slasher 577; d. Rose 2nd 1143 by Longham 104; gr d. Daisy 152 by Necton 3rd 122. See Rose—N 4 by Necton Prize 120.

2512 ROSY — I 13.

Calved May 11, 1881; breeder Mr. R. E. Lofft; owner Mr. W. A. T. Amherst, M.P.; s. Stout 581; d. Rosebud—I 13.

2513 ROSY — R 8.

Calved November, 1879; breeder Mr. T. Easter; owner Mr. Watson; s. Charles 639; d. Dingo 2151 by Read 385; gr d. Beauty—R 8 by a Laxfield Bull.

2514 ROSY MORN — P 3.

Calved November 2, 1882; breeder Mr. J. J. Colman, M.P.; s. Roundhead 564; d. Rosa 1133 by Norfolk Duke 127; gr d. Rosa 3rd 480 by Young Duke 234. See Rose 2nd—P 3 by Tenant Farmer 213.

2515 ROWLEY — V 13.

Calved December, 1882; breeder Mr. J. M. Spinks; s. Lord George 520; d. Lady Rowley 985 by Monarch 241; gr d. Rowley—V 13 by Bullfinch 239.

2516 RUBBISH — O 3.

Calved March 30, 1882; breeder Mr. A. Taylor; s. Starston Duke 570; d. Summer Rose 2559 by Harold 83; gr d. Cossett 1405 by Rifleman 175. See Cowslip—O 3 by Bowbearer 22.

2517 RUBY — K 16.

Calved December 21, 1882; breeder Mr. B. Stimpson; s. Robin Hood 394; d. Rosie 2nd 1811 by Robin Hood 394; gr d. Rosie 1163 by Powell 143. See Cherry Red—K 16.

2518 RUBY A — O 13.

Calved October 3, 1881; breeder the Right Hon. Lord Henniker; s. Cyprus 473; d. Ruby 1165 by Ruddy 402; gr d. Strawberry—O 13.

2520 RUPEE — E 2.

Calved January 20, 1881; breeder Mr. T. Brown; s. Priam 373; d. Rosamond 1136 by Norfolk Duke 127; gr d. Rosebud 489 by Stoke Duke 209. See Rose of Eaton—E 2 by Cringleford Sire 44.

2521 RUSTIC — V 13.

Calved December, 1881; breeder Mr. J. M. Spinks; s. Lord George 520; d. Ruby 1827 by Damian 282; gr d. Lady Rowley 985 by Monarch 241. See Rowley—V 13 by Bullfinch 239.

2522 RUTH — K 26.

Calved 1882; breeder Mr. P. Blofield; s. Rollick 558; d. Rose Princess 2509 by Davyson 3rd 48; gr d. Rosette 1162 by Young Major 235. See Rose—K 26 by a Stoke Bull.

2523 RUTH — V 16.

Calved February 24, 1882; breeder Col. W. Beeston Long; s. Cookley Lad 471; d. Ruby 1167 by Prince 375; gr d. Nancy—V 16 by Hero 322.

2524 SALLY SENTER — A 26.

Calved November, 1881; breeder Mr. Senter; owner Mr. T. Fulcher; s. Lofty 515; d. Suitor—A 26.

2525   SATIN — F 4.

Calved January 21, 1881; breeder Mr. B. Stimpson; s. Robin Hood 393; d. Silky 1190 by Rufus 189; gr d. Snelling—F 4.

2526   SATINETTE — T 7.

Calved January 18, 1882; breeder Mr. W. A. T. Amherst, M.P.; s. Davyson 3rd 48; d. Satin 1837 by Robin Hood 394; gr d. Songster 1859 by Duke of Norfolk 295. See Stranger—T 7.

2527   SEMIRAMIS — N 5.

Calved July 11, 1881; breeder Mr. R. H. Mason; s. Sir Theophilus 576; d. Sheba 1841 by King Cole 330; gr d. Sultana 1876 by Lord Easton 105. See Rose—N 5 by Prince Charlie 151.

2528   SHOTOVER — V 2.

Calved May 24, 1882; breeder Mr. C. Austin; s. Shylock 571; d. Flageolet 1514 by Troston 2nd 590; gr d. Flora 2nd 897 by Doncaster 50. See Flora—V 2 by King Alfred 96.

2529
SHOULDHAM PRIMROSE 15TH — A 1.

Calved June 15, 1881; breeder the Rev. Dr. Allen; s. Priam 373; d. Shouldham Primrose 5th 1181 by The Beau 16; gr d. Shouldham Primrose 3rd 537 by Tenant Farmer 213. See Shouldham Primrose—A 1 by Hero of Newcastle 85.

2530
SHOULDHAM PRIMROSE 16TH — A 1.

Calved July 10, 1882; breeder the Rev. Dr. Allen; s. Priam 373; d. Shouldham Primrose 5th 1181 by The Beau 16; gr d. Shouldham Primrose 3rd 537 by Tenant Farmer 213. See Shouldham Primrose—A 1 by Hero of Newcastle 85.

2531
SHOULDHAM PRIMROSE 17TH — A 1.

Calved September 9, 1882; breeder the Rev. Dr. Allen; s. Priam 373; d. Shouldham Primrose 11th 1845 by Norfolk Duke 127; gr d. Shouldham Primrose 3rd 537 by Tenant Farmer 213. See Shouldham Primrose—A 1 by Hero of Newcastle 85.

2532
SHOULDHAM PRIMROSE 18TH — A 1.

Calved September 14, 1882; breeder the Rev. Dr. Allen; s. Priam 373; d. Shouldham Primrose 13th 1847 by Norfolk Duke 127; gr d. Shouldham Primrose 6th 1182 by Powell 143. See Shouldham Primrose 3rd—A 1 by Tenant Farmer 213.

2533
SHOULDHAM PRIMROSE 19TH — A 1.

Calved November 18, 1882; breeder the Rev. Dr. Allen; s. Priam 373; d. Shouldham Primrose 12th 1846 by Norfolk Duke 127; gr d. Shouldham Primrose 5th 1181 by The Beau 16. See Shouldham Primrose 3rd—A 1 by Tenant Farmer 213.

2534
SHOULDHAM STRAWBERRY 17TH — M 6.

Calved July 20, 1881; breeder the Rev. Dr. Allen; s. Priam 373; d. Shouldham Strawberry 12th 1188 by Royal Duke 181; gr d. Shouldham Strawberry 3rd 541 by Tenant Farmer 213. See Shouldham Strawberry—M 6 by Hero of Newcastle 85.

2535
SHOULDHAM STRAWBERRY 18TH — M 6.

Calved September 27, 1882; breeder the Rev. Dr. Allen; s. Priam 373; d. Shouldham Strawberry 13th 1849 by Norfolk Duke 127; gr d. Shouldham Strawberry 10th 1186 by The Beau 16. See Shouldham Strawberry 3rd—M 6 by Tenant Farmer 213.

2536   SILENT BEAUTY — O 9.

Calved December 30, 1882; breeder Mr. J. J. Colman, M.P.; s. King Charles 329; d. Silent Lass 1189 by Powell 143; gr d. Silence 548 by Rifleman 175. See Silence—O 9.

2537   SILENT WOMAN — O 9.

Calved December 25, 1881; breeder Mr. J. J. Colman, M.P.; s. Rufus 188; d. Silent Lass 1189 by Powell 143; gr d. Silence 548 by Rifleman 175. See Silence—O 9.

2538   SILVER-LOCKS 2ND — B 10.

Calved April 6, 1882; breeder Mr. R. E. Lofft; s. Wild Robin 290; d. Silver-locks 551 by The Baron 10; gr d. Silverbury 550 by Playford Sire 142. See Bury—B 10.

2539   SISTER ANNE — B 12.

Calved February 11, 1882; breeder Mr. H. Biddell; s. Blue Beard 625; d. Shining Hours 1842 by Monarch 4th 351; gr d. Busy Bee 733 by The Baron 10. See The Bee—B 12 by Seneca 195.

2540   SNOWDROP — W 3.

Calved November 17, 1881; breeder Mr. R. E. Lofft; owner the Right Hon. W. H. Smith, M.P.; s. Stout 581; d. Newbourn Pride 6th 1707 by Donald 291; gr d. Newbourn Pride 2nd 384 by Glatton 79. See Newbourn Pride—W 3 by Garibaldi 73.

2541   SONSIE OF ELMHAM — T 7.

Calved June 20, 1882; breeder Mr. Fulcher; s. Lofty 515; d. Songster 1859 by Duke of Norfolk 295; gr d. Stranger—T 7.

2542   SOPHIA — O 13.

Calved June 8, 1881; breeder the Right Hon. Lord Henniker; owner Mr. G. F. Taber; s. Cyprus 473; d. Strawberry—O 13.

2543   SOPHIA — R 1.

Calved April, 1876; breeder Mr. H. Birkbeck; owner Mr. Garrett Taylor; s. Trimmer 218; d. Sweetmeat 594 by Young Duke 234; gr d. Susan 586 by Tommy 216. See Sarah—R 1 by Elmham 65.

2544   SPINSTER 2ND — AA 13.

Calved June 11, 1882; breeder Mr. G. F. Taber; owner Mr. D. L. Stevens; s. Lofty 515; d. Spinster 1861 by Brutus 269; gr d. Sprite 1203 by Rufus 188. See Spot—A 13.

2545   SPOTLESS — A 13.

Calved December, 1881; breeder Mr. J. Howling; owner Mr. G. F. Taber; s. Lofty 515; d. Spot—A 13.

REGISTER OF COWS.

2546   SPRIGHTLY — A 13.
Calved 1878; breeder Mr. J. Howling; s. Brutus 269; d. Spot—A 18.

2547   SPRIGHTLY — R 8.
Calved March 20, 1881; breeder Mr. W. B. Easter; s. Brundish Prince 462; d. Beauty 696 by Read 385; gr d. Beauty—R 8 by a Laxfield Bull.

2548   SPRIGHTLY — 2 Suff.
Calved 1881; breeder Mr. E. Boon; s. Troston 3rd 591; d. Nancy by a Kettleburgh (Turner's) Bull. See Introduction.

2549   SPRING-LEAF — B 21.
Calved January, 1881; breeder Mr. A. J. Smith; s. Pickwick 720; d. Summer-leaf 2558 by Iron Duke 125; gr d. Roseleaf 498 by Seneca 195. See Rosebud—B 21.

2550   STARLING — N 4.
Calved July 26, 1882; breeder Mr. R. H. Mason; s. Philip 538; d. Rosalind 1787 by King Harry 332; gr d. Rose 2nd 1143 by Longham 104. See Daisy—N 4 by Necton 3rd 122.

2551   STOCKDOVE — I 2.
Calved August 21, 1881; breeder Mr. H. le Strange; s. Goshawk 497; d. Ringdove 1782 by Royal Duke 181; gr d. Janet 958 by Arthur 4. See Jenny—I 2 by The Peer 139.

2552   STRAWBERRY 2ND — N 4.
Calved October, 1879; breeder Mr. R. H. Mason; s. King Harry 332; d. Strawberry 1872 by King Tom 335; gr d. Daisy 152 by Necton 3rd 122. See Rose—N 4 by Necton Prize 120.

2553   STRAWBERRY K — O 13.
Calved May 2, 1882; breeder the Right Hon. Lord Henniker; s. Cyprus 473; d. Strawberry 2nd 1212 by Eclipse 2nd 299; gr d. Strawberry—O 13.

2554   STRAWBERRY — R 11.
Calved July 17, 1882; breeder Mr. W. B. Easter; owner Mr. J. Boggis, Jun.; s. Brundish Prince 462; d. Strawberry 1878 by Simon 408; gr d. Pretty 1092 by Read 385. See Pretty—R 11 by a Laxfield Bull.

2555 STRAWBERRY — 2 SUFF.
Calved 1881; breeder Mr. E. Boon; s. Troston 3rd 591; d. Lovely 2330 by a Thornham Bull; gr d. Nancy by a Kettleburgh Bull.

2556 SUFFOLK DUCHESS — B 11.
Calved October 9, 1877; breeder Mr. H. Biddell; owner Mr. A. J. Smith; s. Crown Prince 281; d. Suffolk—B 11.

2557 SUFFOLK COUNTESS — B 11.
Calved March 11, 1881; breeder the Most Hon. the Marquis of Bristol; s. Ironsides 509; d. Suffolk Princess 1875 by Crown Prince 281*; gr d. Suffolk—B 11.

2558 SUMMER-LEAF — B 21.
Calved July 5, 1877; breeder Mr. H. Biddell; owner Mr. A. J. Smith; s. Iron Duke 125; d. Roseleaf 498 by Seneca 195; gr d. Rosebud—B 21.

2559 SUMMER ROSE — O 3.
Calved 1876; breeder Sir E. C. Kerrison, Bart.; owner Mr. A. Taylor; s. Harold 83; d. Cossett 1405 by Rifleman 175; gr d. Cowslip—O 3 by Bowbearer 22.

2560 SUNFLOWER — B 7.
Calved July 20, 1881; breeder Mr. H. Biddell; s. Monarch 4th 351; d. Tiger Lily 1232 by The Baron 10; gr d. Lily 308 by Playford Sire 142. See Grundisburgh Primrose—B 7.

2561 SUPERB — F 4.
Calved January 28, 1882; breeder Mr. B. Stimpson; s. Robin Hood 394; d. Silky 1190 by Rufus 189; gr d. Snelling—F 4.

2562 SUSANNA 2ND — E 12.
Calved March 22, 1882; breeder Mr. J. F. Rogers; s. Emperor 489; d. Susanna 587 by Stoke Duke 209; gr d. Susan—E 12.

2563 SUSIE — I 23.
Calved November 1881; breeder Mr. W. Hudson; s. The Doctor 486; d. Susan 1878 by Quarles Duke 548; gr d. Susan—I 23 by Proud 546.

REGISTER OF COWS.

2564   SUSIE — O 6.

Calved 1880; breeder Sir J. W. C. Hartopp, Bart.; s. Bounty 460; d. Canteen 1361 by a Son of Rifleman 175; gr d. Mirth 344 by Rifleman 175. See Vanity—O 6.

2565   SWEET BRIAR — B 9.

Calved June, 1882; breeder Mr. H. Biddell; owner Mr. W. A. T. Amherst, M.P.; s. The Friar 494; d. Wild Briar 1269 by Iron Duke 125; gr d. June Rose 968 by The Baron 10. See Rose—B 9.

2566   SWEET-HEART — V 5.

Calved December 3, 1882; breeder Mr. H. Biddell; s. Shylock 572; d. White-heart 651 by Seneca 195; gr d. Cherry—V 5.

2567   SWEET PEA — U 14.

Calved December, 1879; breeder Mr. Hunter Rodwell; owner Mr. T. Leonard Palmer; s. Bridegroom 2nd 630; d. Sweet Pea 595 by Waxwork 597; gr d. Sweet Pea—U 14 by Troston Hero 221.

2568   SYBIL 6TH — A 31.

Calved October, 1882; breeder Mr. C. S. Read; s. Haman 499; d. Sybil 4th 1887 by Disraeli 289; gr d. Sylph 1223 by Cherry Duke 32. See Star—A 31.

2569   SYBIL 7TH — A 31.

Calved November, 1882; breeder Mr. C. S. Read; s. Haman 499; d. Sybil 3rd 1886 by Disraeli 289; gr d. Sybil 1223 by Cherry Duke 32. See Star—A 31.

2570   TANSY — U 14.

Calved January 1, 1882; breeder Mr. T. Leonard Palmer; s. Doubtful 487; d. Sweet Pea 2567 by Bridegroom 2nd 630; gr d. Sweet Pea 595 by Waxwork 597. See Sweet Pea—U 14 by Troston Hero 221.

2571   THORNHAM DAVY 3RD — H 1.

Calved December 4, 1882; breeder the Right Hon. Lord Hastings; s. Roscoe 559; d. Thornham Davy 1890 by Thornham Duke 2nd 584; gr d. Davy 16th 845 by Redjacket 7th 169. See Davy 7th—H 1 by Young Duke 234.

2572 THORNHAM PRIZE — K 17.

Calved February 23, 1881; breeder the Right Hon. Lord Henniker; owner Mr. G. F. Taber; s. Cyprus 473; d. Thornham Princess 1230 by Eclipse 2nd 299; gr d. Thursford Queen 599 by Tenant Farmer 213. See Cherry—K 17.

2573 THRIFT — N 2.

Calved November 24, 1881; breeder Mr. T. Leonard Palmer; s. Doubtful 487; d. Lily 4th 1627 by Stout 581; gr d. Lily 3rd 1000 by The Palmer 138. See Lily—N 2 by Hero of Newcastle 85.

2574 TRIMLEY GEM — A 1.

Calved December 31, 1881; breeder Mr. C. K. Cordy; s. Emperor 489; d. Abigail 1301 by May Duke 348; gr d. Lily 998 by Monarch 2nd 114. See Rose of Elmham—A 1 by Redjacket 2nd 164.

2575 TRIMLEY HANDSOME 2ND — X 5.

Calved October 23, 1882; breeder Mr. C. K. Cordy; s. Trimley Tom 589; d. Trimley Handsome 1910 by Blofield 456; gr d. Handsome—X 5 by Zephyr 441.

2576 TROSTON NELLY — A 1.

Calved September, 1880; breeder Mr. R. E. Lofft; owner Mr. W. A. T. Amherst, M.P.; s. Stout 581; d. Elmham Nelly 2nd 1492 by Ronald 397; gr d. Nelly 871 by Hero 2nd 86. See Primrose—A 1 by Elmham Sire 67.

2577 TROUBLESOME — 1 NORF.

Calved 1879; breeder Mr. W. Wiffen; owner Mr. John Baly; s. an Elmham Bull; d. a Pond Cow. See Introduction.

2578 TRUTH — E 2.

Calved February 8, 1881; breeder Mr. T. Brown; s. Priam 373; d. Theresa 1228 by Royal Duke 181; gr d. Tulip 605 by Duke 52. See Cowslip 2nd—E 2 by Spot 206.

2579 TULIP 3RD — W 15.

Calved September 12, 1881; breeder Mr. S. Wolton; d. Perfection 2nd 368; d. Tulip 619 by Duke of Suffolk 58; gr d. Daisy—W 15.

2580   TWIN SISTER A — A 19.

Calved July 28, 1882; breeder the Right Hon. Lord Henniker; s. Cyprus 473; d. Lady Constable—A 19 by a Son of Norfolk Duke 127.

2581   TWIN SISTER B — A 19.

Calved July 28, 1882; breeder the Right Hon. Lord Henniker; s. Cyprus 473; d. Lady Constable—A 19 by a Son of Norfolk Duke 127.

2582   UFFORD BELLE — B 11.

Calved December, 1881; breeder Mr. A. J. Smith; s. Pickwick 720; d. Belle 2010 by Iron Duke 125; gr d. Bellona 705 by The Baron 10. See Suffolk Belle—B 11 by Seneca 195.

2583   UFFORD DUCHESS — B 11.

Calved November, 1881; breeder Mr. A. J. Smith; s. Pickwick 720; d. Suffolk Duchess 2557 by Crown Prince 281; gr d. Suffolk—B 11.

2584   ULTRA — L 3.

Calved March 30, 1879; breeder Mr. J. Margarson; s. Lord of the Manor 338; d. Una 1243 by The Freeman 309; gr d. Upton—L 3 by The Palmer 138.

2585   UNIFORM — L 3.

Calved October, 1881; breeder Mr. J. Margarson; s. Purl 611; d. Una 1243 by The Freeman 309; gr d. Upton—L 3 by The Palmer 138.

2586   UNISON — L 3.

Calved November 5, 1881; breeder Mr. J. Margarson; s. Purl 611; d. Una 1243 by The Freeman 309; gr d. Upton—L 3 by The Palmer 138.

2587   UPLAND — L 3.

Calved March, 1882; breeder Mr. J. Margarson; s. Purl 611; d. Upton 2nd 1921 by Franklin 308; gr d. Upton—L 3 by The Palmer 138.

2588   UPSHOT — L 3.

Calved March, 1882; breeder Mr. J. Margarson; s. Purl 611; d. Upton—L 3 by The Palmer 138.

REGISTER OF COWS.

2589   URBANE — L 3.

Calved July, 1881; breeder Mr. J. Margarson; s. Purl 611; d. Una 2nd 1920 by Lord of the Manor 338; gr d. Una 1243 by The Freeman 309. See Upton—L 8 by The Palmer 138.

2590   VANESSA — A 26.

Calved November, 1879; breeder Mr. J. Fenn; owner Mr. Fulcher; s. Brutus 269; d. Violet 2nd 1925 by The Palmer 138; gr d. Violet—A 26 by Hero 2nd 86.

2591   VENUS 2ND — W 3.

Calved October 20, 1881; breeder Mr. S. Wolton; s. Perfection 2nd 368; d. Vesta 1248 by Oakley 133; gr d. Venus 623 by Duke of Suffolk 57. See Starry—W 3 by Garibaldi 73.

2592   VENUS OF GUIST — A 26.

Calved October, 1880; breeder Mr. J. Fenn; owner Mr. Fulcher; s. Brutus 269; d. Violet—A 26 by Hero 2nd 86.

2593   VERMONT BRIDESMAID — I 9.

Calved July 21, 1881; breeder Mr. R. E. Lofft; owners Col. J. B. Mead and Mr. R. J. Kimball; s. Stout 581; d. Bridesmaid 3rd 722 by Cherry Duke 32; gr d. Bridesmaid—I 9 by Rudham Hero 183.

2594   VERMONT TOPKNOT — W 10.

Calved July 31, 1881; breeder Mr. R. E. Lofft; owners Col. J. B. Mead and Mr. R. J. Kimball; s. Doubtful 487; d. Topknot 3rd 1954 by Bright 267; gr d. Topknot 609 by Duke of Suffolk 57. See Cherry—W 10.

2595   VESTAL — A 26.

Calved 1878; breeder Mr. J. Fenn; owner Mr. Fulcher; s. Brutus 269; d. Violet 2nd 1925 by The Palmer 138; gr d. Violet—A 26 by Hero 2nd 86.

2596   VESUVIUS — W 3.

Calved January 12, 1882; breeder Mr. S. Wolton; s. Perfection 2nd 368; d. Venus 623 by Duke of Suffolk 57; gr d. Starry 563 by Garibaldi 73. See Nelly—W 3 by Robinson 178.

2597   VICTORIA 2ND — O 1.
Calved March, 1881; breeder Sir E. C. Kerrison, Bart.; s. Lofty 515;
d. Victoria 625 by Rifleman 175; gr d. Oakley 398 by Rifleman 175. See
Duchess of Suffolk—O 1.

2598   VICTORIA — W 3.
Calved September 26, 1882; breeder Mr. S. Wolton; s. Wild Rover 605;
d. Venus 623 by Duke of Suffolk 57; gr d. Starry 563 by Garibaldi 73.
See Nelly—W 3 by Robinson 178.

2599   VINAIGRETTE — B 21.
Calved December 18, 1882; breeder Mr. H. Biddell; s. Blue Beard 625;
d. Perfume 1731 by Monarch 4th 351; gr d. Sweet-bloom 1221 by The
Baron 10. See Rose-bloom—B 21 by Seneca 195.

2600   VIOLA — N 6.
Calved October 7, 1882; breeder Mr. R. H. Mason; s. Slasher 577; d.
Violet 4th 1252 by Lord Easton 105; gr d. Dainty 819 by Prince Charlie
151. See Nancy—N 6 by Fransham Captain 71.

2601   VIOLA — N 19.
Calved March 25, 1882; breeder Mr. W. A. T. Amherst, M.P.; s. Karl
512; d. Violet 1253 by Simon 407; gr d. Victoria—N 19 by Elmham Bull.

2602   VISCOUNTESS 3RD — U 48.
Calved September 1, 1881; breeder Mr. W. G. Collins; owner Mr. G. J.
Paine; s. Hunston Duke 505; d. Viscountess 2nd 1928 by Prince 376;
gr d. Viscountess—U 48 by Prince Arthur 150.

2603   WALLFLOWER 2ND — Y 1.
Calved January 21, 1883; breeder Captain J. Borlase Tibbits; s.
Plowboy 540; d. Wallflower 1929 by Young Foxhall 437; gr d. Gilly-
flower 2nd 917 by Prince Leopold 380. See Gillyflower—Y 1.

2604   WAXY — U 9.
Calved January 8, 1882; breeder Mr. W. A. T. Amherst, M.P.; s.
Doubtful 427; d. Waxwork 6th 1932 by Hector 319; gr d. Waxwork 2nd
648 by King of Carlford 100. See Waxwork—U 9.

2605   WIFFEN — 1 NORF.
Calved December, 1880; breeder Mr. J. W. Vincent; owner Mr. Fulcher;
s. Falstaff 303; d. a Pond Cow. See Introduction.

REGISTER OF COWS.

2606 WIFFEN CHERRY — 1 Norf.
Calved 1877; breeder Mr. Wiffen; owner Mr. W. Bradfield; s. an Elmham Bull; d. a Pond Cow. See Introduction.

2607 WILD CHERRY — V 2.
Calved May 20, 1881; breeder Mr. G. Gooderham; s. Troston 3rd 591; d. Wild Rose Cousin 1938 by Troston 424; gr d. Favourite 222 by Doncaster 50. See Flora—V 2 by King Alfred 96.

2608 WILD RHONA — V 1.
Calved November 27, 1882; breeder Mr. G. Gooderham; s. Shylock 571; d. Wild Rose of Kilburn 1939 by Troston 424; gr d. Wild Rose 1271 by The Claimant 34. See Rosy—V 1 by Perfection 140.

2609 WILD ROSY — V 1.
Calved March 10, 1882; breeder Mr. G. Gooderham; s. Shylock 571; d. Wild Rose 1271 by The Claimant 34; gr d. Rosy 513 by Perfection 140. See Beauty—V 1 by Wonder 230.

2610 WINNIE — N 2.
Calved June 14, 1882; breeder Mr. R. E. Lofft; owner the Right Hon. W. H. Smith, M.P.; s. Powerful 728; d. Lily 4th 1627 by Stout 581; gr d. Lily 3rd 1000 by The Palmer 138. See Lily—N 2 by Hero of Newcastle 85.

2611 WISE PRINCESS — B 17.
Calved September 23, 1881; breeder Mr. F. D. Kent; s. Young Prince 608; d. Wisdom 656 by Seneca 195; gr d. Fairy—B 17.

2612 ZENOBIA — N 5.
Calved July 7, 1882; breeder Mr. R. H. Mason; s. Philip 538; d. Sheba 1841 by King Cole 380; gr d. Sultana 1876 by Lord Easton 105. See Rose—N 5 by Prince Charlie 151.

2613 ZILLAH — C 1.
Calved May, 1880; breeder Mr. H. Overman; owner Mr. B. Stimpson; s. a Cranmer Bull; d. by Cranmer Duke 2nd 41; gr d. Ruby 2nd 515 by Cranmer Duke 40. See Ruby—C 1 by Rufus 184.